EVERYTHING YOU NEED TO KNOW

ORGANIC
FARMING

PETER V. FOSSEL

Voyageur
Press

First published in 2007 by MBI Publishing Company and Voyageur Press, an imprint of MBI Publishing Company, 400 1st Avenue North, Suite 300, Minneapolis, MN 55401 USA

Voyageur Press titles are also available at discounts in bulk quantity for industrial or sales-promotional use. For details write to Special Sales Manager at MBI Publishing Company, 400 1st Avenue North, Suite 300, Minneapolis, MN 55401 USA.

To find out more about our books, join us online at www.voyageurpress.com.

On the cover: The author with a typical mid-summer's CSA weekly offering.

On the frontispiece: Hoop house tomatoes, or field tomatoes, for that matter, can be efficiently suspended with the Florida weave system shown here. *Courtesy of Eaton's Creek Organics*

On the title page: *(left to right)* A farm stand should be at least as attractive and tasteful as your produce.

Orange sells best, but carrots come in a host of colors and varieties.

Livestock, in this case Belted Galway cattle from Scotland, must have access to open, organic pasture land under organic certification rules.

Flowers are a good addition to a farmers' market produce booth.

On the back cover: *(top right)* We've never had trouble selling fresh eggs from free-range hens, if only because they're so delicious. Don't be afraid to charge more than the most expensive store eggs because yours are fresher. *(bottom left)* Zinnias are so easy to grow, it's hard to find a reason not to have some at your farm stand or farmers' market booth. *(bottom right)* We mixed purple cabbages with other vegetables and flowers in our display gardens, just for fun.

ISBN-13: 978-0-7603-2469-1

Library of Congress Cataloging-in-Publication Data

Fossel, Peter V.
  Organic farming : everything you need to know / by Peter Fossel.
    p. cm.
  ISBN-13: 978-0-7603-2469-1 (softbound)
  ISBN-10: 0-7603-2469-7 (softbound)
  1. Organic farming. I. Title.
  S605.5.F66 2007
  631.5'84--dc22
                                2006034494

Editor: Jennifer Bennett
Designer: Brenda C. Canales

Printed in Singapore

CONTENTS

Chapter 1

WHY ORGANIC?

The farmstand and upper gardens at Plum Hill Farm were designed to make people want to visit.

I like to smell dirt. On a spring day, a planting day, I like to scoop both hands into the earth I tend, lift it up to my face, and inhale its fragrance. Rich, fertile soil smells of the forest floor, teeming with health and life; with beneficial microbes, bacteria, earthworms, and humus. Organically rich soil is heady, almost magical. It's the Merlin of dirt. It is why I am a grower.

I'll never be so mechanized that I can't smell the dirt and know what it needs. Ever.

Fertile soil is everything. America moved away from that fact in the 1940s when soil began to become little more than a planting medium, with fertility entrusted to petroleum-based chemicals, and pest control entrusted to DDT. But we're coming back now; back to earth as it were. In fact, we're coming back with something of a vengeance.

Organic produce was a $3 billion industry in America in 1996. Today it's a $20 billion industry, growing at 20 percent a year or more—by far the fastest-growing sector in agriculture. Organic croplands in the United States more than doubled between 1992 and 1997, and doubled again for many crops by 2003, according to the USDA's Economic Research Service. The organic poultry and dairy sectors grew even faster. Certified organic cropland and pasture now encompasses 2.3 million acres in 49 states.

The reasons for this are many.

TASTE

One reason was expressed to me a while back by a schoolchild whose class had come to our organic farm on a field trip. I encouraged the kids to sample our edible herbs, flowers, and vegetables, whereupon one youngster snipped off a floret of broccoli, tried it, and then looked up in amazement.

These Fruits and Vegetables have been coated with FDA approved food grade Vegetable-, Petroleum-, Beeswax-, and/or Shellac-based wax or resin to maintain freshness:

apples
avocados
cucumbers
eggplant
grapefruit
lemons
limes
nectarines

oranges
parsnips
passion fruit
peaches
peppers
pineapples
plums
pumpkins

rutabagas
sweet potatoes
squash
tangerines
tangelos
tomatoes
turnips

ORGANIC   ORGANIC   ORGANIC

Every year, more supermarkets open and promote an organic section.

DELVIN FARMS

CERTIFIED
ORGANIC PRODUCE

"OUR BUSINESS IS GROWING"

www.delvinfarms.com

A day at the farmers' market; one of the most enjoyable times you can imagine.

"So *that's* what broccoli's supposed to taste like!" he said.

Indeed, taste is one of the foremost reasons given by consumers who seek out organic vegetables at our booth in the local farmers' market. We have been able to charge $8 a pound for mesclun salad mix, and 50 cents an ear for sweet corn, based on taste alone. Freshness is also a factor, of course, but chefs in the highest-caliber

Antioxidants

Fruits and vegetables grown organically contain higher levels of cancer-fighting antioxidants than conventional food, and the difference is enough to have a "significant impact on health and nutrition," according to a recent study published in the journal of the American Chemical Society, the world's largest scientific society.

The levels were 58.5 percent higher in corn, 50 percent higher in marionberries (a type of blackberry), and 19 percent higher in strawberries. "We expect these results to be transferable to most produce," says Alyson Mitchell, Ph.D., a food scientist at the

University of California, Davis, and lead author of the paper, "and it's definitely changed the way I think about my food."

The antioxidant compounds are produced by plants in response to stress, such as that produced by insect pests or competing weeds. The need for these natural safeguards decreases with the use of herbicides and pesticides in conventional agriculture. "By synthetically protecting the produce from these pests, we decrease their need to produce antioxidants," Mitchell says. "It suggests that maybe we are doing something to our food inadvertently."

Our strawberries almost never failed to sell out.

Selling out of the back of a pickup truck can be both simple and effective. This one has a roof for shade, and provision for a hanging scale.

restaurants also seek out organic vegetables and fruit for taste reasons alone.

Alice Waters, owner of the renowned Chez Panisse restaurant in Berkeley, California, states flatly, "The best-tasting food is organically grown."

A study by the United Nations Food and Agriculture Organization found that organically grown Golden Delicious apples received higher taste scores than conventional apples, and another recent study reached the same conclusion for tomatoes and carrots.

"Organic ingredients are becoming a point of pride for chefs in every segment (of the food service industry)," concluded a 2005 survey by *Restaurants and Institutions* trade magazine.

Taste is also a factor in the growth of organic livestock and poultry operations. Those who have tried organic beef,

poultry, or eggs (including myself) are willing to pay a premium for the product. Under federal rules enacted in 2002, organic animals cannot be given antibiotics, growth hormones, or feed made from animal byproducts, which can transmit mad cow disease. No genetic modification or irradiation is allowed, nor is field fertilizer made from sewage sludge or synthetic ingredients.

ECONOMICS

Another reason for growing organically is profitability. Organic growers not only eliminate the cost of fertilizers, pesticides, fungicides, and herbicides, but can produce yields at least comparable to conventional (chemical) methods, and often greater, while selling their products at higher prices.

"We were using chemicals I don't even want to tell you about, and it seemed like

This is a Saturday morning market in Franklin, Tennessee. At the farmers' market, display your produce in abundance. When a basket or container is half-empty, change to a smaller basket (the hardest potato to sell is the last one).

we needed more every year. The cost was killing me," says Hank Delvin, a farmer in College Grove, Tennessee, who switched from conventional to organic methods 10 years ago. Delvin now uses cover crops, compost, crop rotation, and some organic fertilizer to feed the soil on his 60-acre farm, which only has about 40 acres under cultivation at one time. The land is more productive than ever, he says. The Delvin farm has put three kids through college, and their community-supported agriculture (CSA) is now the most successful in the region.

My own experience was similar. After moving from land I'd owned and used chemicals on for nearly 20 years, we found a former horse farm that had decades' worth of horse manure composting over a bank on the east pasture. We moved this composted manure and bedding to where we wanted it, and our produce yields began to skyrocket. We never used or needed that much composted manure again, but brought in leaves and untreated grass clippings from local landscape contractors and combined that with cover crops and other ingredients to produce the most fertile and productive soil I've ever known.

Recent studies bear this out scientifically.

"Organic farming produces the same yields of corn and soybeans as does conventional farming, but uses less fossil energy, less water, and no pesticides," concluded a 2005 farming trial that tracked costs and benefits of both farm types for 22 years at Cornell University. Moreover, in the drought years of 1988 to 1998, corn yields within the organic system were 22 percent

Land used for hay, such as this behind our house, usually can be certified organic immediately.

higher than conventional yields, in great part because soils rich in organic matter hold moisture better than conventional farmland.

"Organic farming offers real advantages," said David Pimentel, a Cornell professor of ecology and agriculture and lead author of the study.

While corn yields were about one-third lower during the first four years of the study, Pimentel said that over time the organic systems produced higher yields—again, especially under drought conditions.

"Organic approaches . . . not only use an average of 30 percent less fossil energy, but conserve more water in the soil, induce less erosion, maintain soil quality, and conserve more biological resources than conventional farming," he added.

While labor costs were about 15 percent higher on the organic farms, the higher prices that organics command in the marketplace make the net return equal to or higher than conventionally produced crops, Pimentel said.

A recent 21-year study in Sweden mirrored the one at Cornell. The Swedish study, published in a 2006 issue of *Science* magazine, reported that while organic methods can result in lower crop yields, especially in the first few years, they used 50 percent less energy, 97 percent fewer pesticides, and more than 50 percent less fertilizer than chemical methods, making organics highly viable as a farming system.

"To get 90 percent of conventional yields using only half the fertilizer is incredible," noted one academic observer.

A 2001 study done at Washington State University and published in the journal, *Nature*, concluded that organic apple farming was also not only better for the soil and the environment, but resulted in comparable yields, higher profits, and greater energy efficiency.

One reason it takes several years for organic methods to reach yields comparable to chemical farming is that chemicals degrade or eliminate the rich populations of microbes, bacteria, and fungi so vital to organically fertile soil. This array of microscopic life, often called the "microherd," is what converts organic matter into plant nutrients, and once the microbe population is degraded it takes time to build it up again.

OUR HEALTH

Even more so than taste, health is the reason most often given by our customers for buying organically. They don't want to ingest through their food chemical herbicides, fungicides, or the 1.2 billion pounds of pesticides released legally in the United States every year.

Mind you, government regulatory agencies and most of the food industry tell us that our food is safe to eat, and in great part it probably is. Probably. But there are doubts, and these doubts are driving consumers and growers alike into organics.

An example of a wildly popular commercial herbicide, Roundup. To control weeds and aquatic plants, Roundup (glyphosate) and its derivatives have been approved by the EPA for use on ponds, recreational waterways, reservoirs, and wildlife sanctuaries. Roundup's manufacturer, Monsanto, has maintained that the herbicide affects only targeted plant species, and poses little threat to aquatic or terrestrial wildlife. However, recent research at the University of Pittsburgh paints another picture.

Rick Relyea, an assistant biology professor at the university, found that Roundup is "extremely lethal" to amphibians. His study, published in the April 2005 issue of Ecological Applications, found that an aquatic application of Roundup caused a 22 percent decrease in aquatic species richness, a 70 percent decrease in amphibian biodiversity, and posed many health threats to tadpoles. The herbicide completely eliminated populations of Leopard frog and wood frog tadpoles, while almost eliminating all toad tadpoles. Those that survived experienced an 86 percent decline in body mass.

Kale is a highly nutritious spring crop sold in bunches here at a farmers' market.

One doubt stems from the fact that, by federal law, no pesticide can be marketed as "safe." Pesticides are toxins, and toxins are inherently deadly in certain doses. "Risk" is the appropriate word when a pesticide is deemed acceptable by the Environmental Protection Agency (EPA). To quote the EPA, "In many cases, the amount of pesticides people are likely to be exposed to (through foodstuffs) is too small to pose a risk." Personally, I dislike the use of the words "many" and "likely" in that context.

The EPA, in a paper on human health issues and pesticides, goes on to report that "some (pesticides), such as the organophosphates and carbamates, affect the nervous system. Others may be carcinogens. Others may affect the hormone or endocrine system in the body."

It's long been thought that pesticides don't cross the placenta to a fetus in the womb, but a 2005 study by the American Red Cross tested 21 common pesticides, and found that all 21 made the crossover.

French scientists reported in January 2006 that exposure to pesticides in the womb or as a child can double the risk of developing acute leukemia. Youngsters exposed to fungicides and garden insecticides had more than double the rise of the illness than other children, the researchers determined. Even insecticidal shampoos used to kill head lice increase the odds of the disease.

A recent University of Washington study found that children eating organic fruits and vegetables had concentrations of organophosphates (OP) in their urine six times lower than children eating conventional produce. An earlier study cited by the authors looked at pesticide residue in the urine of 96 children, and found OP pesticide residue in all but one. Parents of that child reported eating only organic produce.

Eggs are a huge seller at our farmers' market; everyone sells out each week.

The studies focused on children because pesticide exposure puts them at greater risk for two reasons: They eat more food relative to body mass, and, being smaller than adults, equal doses of chemicals have greater effect.

Others are concerned that "safe thresholds" for certain toxins are continuously revised downward as more evidence comes in, as was true for DDT and lead. The initial "safe" blood level for lead in 1960 was revised down drastically in 1990, and new studies suggest there is no safe threshold for lead or mercury.

"Such results raise serious questions about the adequacy of the current regulatory (system), which by design permits children to be exposed up to 'toxic thresholds' that rapidly become obsolete," says the Greater Boston Physicians for Social Responsibility. The group goes on to say that combinations of commonly used pesticides can multiply the toxic effect.

USDA data cited in a 2006 *Consumer Reports* study indicated that the following fruits and vegetables contained the highest pesticide levels even after washing: apples, bell peppers, celery, cherries, spinach, winter squash, green beans, strawberries, peaches, pears, potatoes, and red raspberries.

The good news is that those of you reading this book are likely to have the desire and wherewithal to grow not only your own chemical-free produce, but to grow it for others.

NUTRITION

Recent studies in professional, peer-reviewed journals have shown organic foods to have higher nutritional value than conventional foods. For example, researchers at the University of California, Davis, also recently found that organic tomatoes had higher levels of phytochemicals and vitamin C than conventional tomatoes. (Phytochemicals work alone and in combination with vitamins and other nutrients in food to prevent, halt, or lessen cancer, heart disease, and other diseases.)

Nutrition Australia, an Australian research foundation, has determined that "The best . . . studies show consistency in higher vitamin C (and perhaps iron and magnesium) [levels] in organic food, and consistently less unwanted and potentially toxic nitrates. Levels of antioxidants and some other health-related food components

More than 200 years ago, in 1793, a Virginia woman named Martha Randolph wrote to her father in Washington, complaining of insect damage in their garden. Her father's advice was this:

"We will try this winter to cover our garden with a heavy coating of manure. When earth is rich, it bids defiance to droughts, yields in abundance, and of the best quality. I suspect the insects which have harassed you have been encouraged by the feebleness of your plants, and that results from the lean state of your soil."

The father was Thomas Jefferson, one of the foremost horticulturists in American history.

Two keys to a good market day are: a genuine smile and a memory for names.

have generally been reported to be higher in organic foods."

Organics are even richer in salicylic acid, which is responsible for the antiinflammatory action of aspirin and has been shown to reduce hardening of the arteries and bowel cancer. This is from a 2002 United Kingdom study of organic and conventional soups, in which the organic brands contained almost six times as much salicylic acid as the conventional soups.

More studies are under way, and much research remains to be done—but the straws are in the wind, and it's not hard to tell which way they're blowing.

EARTH-FRIENDLINESS

In the Organic Trade Association's list of the 10 best reasons to go organic, more than half have to do with the environment—with cleaner water supplies, more fertile soil, more sustainable farm practices, greater respect for a balanced ecosystem, and an increased effort among many organic growers to preserve our heirloom seed bank and promote biodiversity.

Organic-growing methods also tend to be friendlier to wildlife, birds, amphibians, earthworms, bees, and other beneficial insects, in part by eliminating reliance on

chemical pesticides, herbicides, fungicides, and fertilizers. These often not only eliminate organisms, but degrade or eliminate their habitat.

Organic farming may also have a significant impact on slowing global warming, in that soils rich in organic matter absorb and retain significant amounts of carbon, according to the 2005 Cornell study comparing the economics of organic and conventional farms. The organic farm had 15 to 28 percent more carbon, which was the equivalent of taking about 3,500 pounds of carbon dioxide per hectare out of the air.

Environmental costs of chemical farming include erosion, water pollution, and damage to wildlife biodiversity. A recent study by Iowa State University economists determined that these "external" costs of American agriculture amount to between $5 billion and $16 billion per year. If these were accounted for in the price of conventional produce, then the retail price difference between organic and conventional products would be narrowed or eliminated. And the study did not consider health costs of chemical agriculture.

TO MARKET, TO MARKET

In selling almost anything successfully over time, no matter what it is, quality is everything. Marketing can also be everything. You not only need a superior product, but need to get it into the hands of those who want it.

Like all growers, many organic farmers find marketing is the hardest part of farming. Some lack the skills and creativity to find profitable outlets; some simply dislike dealing with the public, doing research, or addressing the other details essential to successful marketing.

Still, the work isn't difficult, and the demand for organics has made it a seller's market. When we first went organic, we found several local restaurants and small markets eager to buy our produce at a premium. Our farm and nursery were designed to be as beautiful and user-friendly as possible to boost business at our farm stand. The *Boston Sunday Globe* called our farm "a lush living painting," and that did more to boost sales than anything I could dream up. Many came to wander through the market gardens and relax—and then they bought. That was the whole idea.

I also wrote a free gardening column in the local weekly newspaper in exchange for free advertising space. We invited school groups out to see the farm, and helped local schools begin their own gardens.

Many organic growers offer newsletters, either via e-mail or regular mail, and others boost sales with value-added steps. These can range from offering washed greens to offering recipes for cooking kale.

Farmers' markets have been booming in recent years, and these make an ideal outlet for organic growers to sell direct to consumers. You may not be able to charge much of a premium, but then again you might—and it's generally better than selling wholesale. If nothing else, you'll know the price you get is fair, and not subject to whims of the national or international marketplace. Consumers flock to farmers' markets because they not only can find lower-priced organic food, but they know the food is fresh and they're supporting local agriculture. If your local area doesn't have a farmers' market, it's easy enough to start one. More on that in Chapter 12.

Another boon to organic growers has been the recent growth in community-supported agriculture (or CSAs, as they're called). The idea is to pair local consumers who want fresh organic produce from a local farm with farmers who want a stable local market. Typically a family will pay in advance for a known quantity, or share, of the farmer's produce for that year.

A farming friend of ours offers a half-bushel of certified organic produce at $25 per week, or $700 for the 28-week season. The $700 is paid in advance, or can be paid in two increments for slightly more. They also offer a half-bushel of produce every other week for $375 per season. The baskets (boxes, actually) can be picked up at the farm, or at three other points, including the local farmers' market on Saturdays. They now have about 220 customers, or share-holders. Do the math, and you'll see how attractive the system is to all concerned.

SO WHAT ABOUT PESTS?

I'm often asked what organic farmers do about insect pests, and the answer is very little. Why? Because pests are predators who go after the stressed, the weak, the infirm. Vegetables growing in fertile soil rich with organic matter have access to all the nutrients and minerals they need, and therefore are extremely healthy—and therefore immune to much of the insect damage so prevalent on chemical farms.

In a word, don't feed your plants, feed your soil. Enrich your soil, and the plants will take care of themselves.

I've seen this proven time and time again. Most recently it happened when we opened a new 50-foot greenhouse and went to the state department of agriculture for a greenhouse inspection certificate. I was asked if I had a pesticide applicator's license, and I explained that, well, no, I didn't have one because I don't use pesticides.

"You don't use *pesticides*? How can you run a commercial greenhouse without pesticides?" The fellow was clearly dumbfounded.

I explained that we fed our plants a mix of compost tea and a mix of fish and kelp emulsion. This left us with plants so healthy that pests and disease didn't bother them.

This notion sort of ricocheted off his mind, so we agreed to talk again in five months to see how things were going with the roughly 10,000 seedlings I expected to raise.

At the appointed time I called the fellow and he asked how many of my 10,000 plants I'd lost.

"Six," said I.

"Only SIX?!" said he.

"Yep. Six. Usually we do better."

He still thought I was a bit out there in my thinking and ways, but the fact is that so-called "conventional" chemical methods have been around for less than 75 years, where organics have been employed for 10,000 years of recorded agriculture. And my thoughts about feeding the soil, not the plant, are nothing new at all.

A summer CSA box from Tennessee's Delvin Farms.

In 1910, grains and cereals made up 36 percent of the protein in the average American's diet. Today the figure has dropped to 17 percent.

THE NEW FARM

Speed is everything now; just jump on the tractor and way across the field as if it's a dirt-track. You see it when a new farmer takes over a new farm: he goes in and plants straight-away, right out of the book. But if one of the old farmers took a new farm, and you walked round the land with him and asked him: "What are you going to plant here and here?" he'd look at you some queer; because he wouldn't plant nothing much at first. He'd wait a bit and see what the land was like: he'd *prove* the land first. A good practical man would hold on for a few weeks, and get the feel of the land under his feet. He'd walk on it and feel it through his boots and see if it was in good heart, before he planted anything; he'd sow only when he knew what the land was fit for.

—*Wendell Berry, The Gift of Good Land*

The upper, roadside gardens at Plum Hill Farm were all designed to attract customers for repeat purchases.

One of the happiest moments of my life was the day I lost my publishing job. I'd been editor of *Country Journal* magazine for years, but the parent company was positioning its magazines for sale and so let its higher-paid people go to improve the profit margin.

What I lost was a lot of stress, and what I returned to was the farm—with its horses, chickens, heifers, and market gardens we'd been growing for years. There, my first question to the universe was, "What now?"

The answer came back in a knowing, a sort of voice without words, and the knowing said to me, "Farm."

I said to myself, "That's insane. I'm an organic gardener, not a farmer."

And the knowing replied, "Then plant a VERY BIG garden."

So we did.

We were blessed in the enterprise because we'd been improving the soil for years, before we had to depend on it for income. We'd been selling for years also, so we knew planning, pricing, labeling, markets, and the other crucial details that have little to do with seeds and soil. To improve the new acres we'd be planting, we called on local landscape contractors to bring us their leaves and untreated grass clippings. We had plenty of manure—and living near the ocean, we had unlimited access to seaweed, and knew we'd not be lacking for any nutrients. So many leaves and grass clippings arrived that we had to turn and stir them with a front-end loader.

We erected two greenhouses (one of them heated), along with a rustic, timber-framed roadside stand. I wrote a free weekly column on organic growing for the local newspaper in exchange for free weekly ad space announcing the week's harvest. We visited local restaurants and nailed down agreements on mesclun mix and dinner vegetables, and did the same with the local fish market.

We used the heated greenhouse as a nursery for transplants, and then set about making our nursery, roadside gardens, and roadside stand just as beautiful and alluring as we could make them with the thought in mind that people would stop just to browse around, *then* maybe buy.

And buy they did. They stopped; they bought; they were hooked; they told friends, and they came back. Then the

Boston Sunday Globe did a big article on our enterprise, announcing that "a trip to Plum Hill Farm is like stepping into a lush, living painting."

Yes, sir, that was the tipping point. But it was coming anyway. School groups had been dropping by on field trips, and Janet and I had helped plan and sow organic gardens for area high schools to help spread the word on organics and fresh produce. People said our produce tasted like nothing they'd ever tasted before, but they also bought herbs from us, and cut flowers. We always had big bouquets at the open farm stand, luring one car in after another off Union Street. We left no trick unturned to get that reluctant old 747 of a farm off the ground and up to cruising altitude.

We were able to do that because we already had a house, barn (optional), and land; we had years of experience growing organically for the market; we started, and remained, small; our pleasures and needs were simple; and we didn't borrow money to get the farm going. All of these are key to a new farm, but the last is particularly important and runs totally opposite of most economic thinking.

This schooner weather vane made a nice focal point in the upper gardens.

Automatic overhead sprinkler heads make life easier in the hoop house. I prefer hand-watering, probably because I've always done it that way but also because it forces me to check all my flats for problems every day.

In America, when consumers are confident—as evidenced by increased consumer spending and borrowing—the national economy is said to be healthy. As it happens, what's true for America isn't true for the farmer.

I don't know how they balance checkbooks in Washington, but every time I increase spending and borrowing around our place the household economy goes straight to hell. Mind you, banks need our loan interest to thrive and grow, just as corporate manufacturers need us to buy their latest products, but a certain comfort and sense of independence comes with saving, not borrowing, for one's needs. If these needs are simple and fail to bolster the national economy, then all we can do is hope the government will muddle through without our help a while longer.

The problem lies in trying to preserve, or even foster, a pastoral agrarian life in the midst of an industrial economy. The two don't mix well. Industry thrives on expansion, for one thing, and for this it needs capital. It needs to borrow—thence to produce, sell, grow, and earn profits at a rate greater than interest charges. When a farmer tries this—borrowing, expanding, amassing capital, however, he falls flat on his face.

Oh, a fair living can be made here and there (for a while perhaps), but food cannot be produced and sold at a profit margin greater than the interest rates on commercial bank loans. The more milk or lettuce that's produced in a market, the lower the price, and you can't come out with a "new, improved model" each year, or create new demand through advertising.

You can, however, gain comfort from the fact that people must eat, and they prefer to eat the best they can find. So don't do it the way corporate America does it. Instead, focus on quality, efficiency, reputation, barter, social supports, and a sense of community. Make your farm beautiful. Love the soil. Know your plants. Grow for taste, not shelf life, looks, or whether your tomatoes will bruise on the trip from California. Grow for life and health.

You'll get up earlier than anyone you know, and finish work later. But you'll have one of the lowest-stress occupations around. And, you'll take advantage of the fact that two of the best natural cures for depression are exercise and sunlight. You won't be depressed.

Getting back to new farms, the first key, again, is to try growing on land you already own.

Every new farmer's experience will be different, but the lessons from our own years at Plum Hill Farm, and from when we moved to a new magazine job (and a new, start-all-over-again-farm) in Tennessee, all point to a few tricks to starting an organic farm successfully. The first is simple.

START SMALL

If 50 acres seems small, then start smaller. I know experienced growers who make a nice living on 5 acres or less. My friend Eliot Coleman cultivates less than 3 acres on his farm in Harborside, Maine. I've done the same myself. Our experience is that two adults and one or two interns (or children) can manage no more than 5 acres of farmland if they are to produce only top-quality organic produce.

The income one can earn off an acre of land varies widely with the soil, climate, market, expertise of the grower, and other factors, but if you were to grow 80 cents'

Cosmos are easy to grow and make an excellent cut flower—and cut flowers are an excellent draw to a farmstand or farmer's market booth.

This field of potatoes is part of an intensely-planted farm on less than three acres. *Courtesy of Eaton's Creek Organics*

worth of crop per square foot, then one acre of land could realize a gross income of about $34,000. That's *gross*, mind you, but it isn't bad. And I know of people who have done even better than that.

We calculated once that on 3 acres in Massachusetts we grew enough vegetables to supply 50 to 75 people for a year. If we hadn't sold herbs, cut flowers, and nursery stock, the amount of produce grown could have been even greater. And we could have cultivated another 2 acres or so if we hadn't been working part-time off the farm, which means it wouldn't have been difficult to grow enough for 100 people or more. When demand became extremely high for one or another crop, as it did for our mesclun mix, rather than grow more, we raised the price until supply and demand met again. We ended up charging more for this than you'd pay for the finest steak.

One advantage of staying small is the need for less equipment. We got by with a tiller, seeder, and a pickup truck. Whatever else we needed we rented or bartered the use of from neighbors. Since then I've learned the value of a small tractor and attachments, but we did without them once and probably could again.

Rather than strive for growth in terms of acreage, we grew by constantly improving the quality of our crops and the service we provided. People came to our farm stand, and bought from us at the farmers' market, not because we had a big farm but because we offered what they can't find at the supermarket—freshness, flavor, variety, service, and a sense of community. We helped others learn how to grow their own, and offered free recipes for our products. In time, I believe we helped to hook an entire town on cooking with herbs.

Our farm and market gardens were also open behind the scenes to the public. We

Blackberries must be pruned back every year. Note the rabbit.

When buying farmland, be sure the water supply is not only present, but sufficient, as with the springs that feed this creek.

placed chairs around so people could sit among the herbs, flowers, and vegetables. Many said they gained a sense of peace from coming to the farm. One woman came by every day for days, explaining finally that she'd just lost her husband, and our plant beds were a place she found solace. I know the feeling.

SOME GOOD LAND

Buying farmland is tricky under the best of circumstances, and organic farming gives it an additional twist because the soil, not Monsanto, will be feeding your crops. So you want the land to be the best you can find. Buy the richest land you can afford, and no more of it than you need.

In making any decision, consider the following:

1. The market. Whatever you grow will have to sell somewhere, so in evaluating land, consider whether the road it's on is busy enough to support a roadside farm stand, or is near enough to an existing farmers' market. If it's not, then you can start up a farmers' market (see Chapter 13), but this is a complication you may not want.

If you're considering setting up a CSA, is there sufficient population nearby to accommodate such a venture? And would the local populace support a new organic farm? Organics are a much easier sell in some areas, such as college towns, than in others. Ask around, see what the competition is (if any). Talk with other farmers and store managers in the area. Find out if enough restaurants would want your produce to make that market worthwhile.

Transportation costs to and from your market can seriously eat into profits given today's gas prices. A farmer's time is a most valuable resource, and none of us wants to spend much of it in a truck on the interstate.

Buy the richest land you can afford, and no more of it than you need.

Do this test: When the soil is just damp, dig a hole 8 inches deep and bring out a clump the size of a tennis ball. Squeeze it slightly to see if it holds its shape. Soil rich in organic matter has various-sized lumps that hold their shapes well. Overly sandy soil will fall apart, and heavy clay soil will have clumps difficult to crumble apart.

At the same time look for earthworms, beetles, centipedes, and other signs of life. All of these indicate healthy soil.

A small-wheeled scuffle hoe is priceless for cultivating small acreage.

At the same time, inexpensive farmland is rarely found cheek-by-jowl to an upscale market. Few can afford to buy farmland (or even live) in Boulder, Colorado, for example, but Boulder has one of the finest and liveliest farmers' markets I've ever seen, and most of the produce is organic. The same is true of our own farmers' market in Franklin, Tennessee. The town is one of the prettiest nineteenth-century towns in America, home to many wealthy folks, and it too has a booming farmers' market, with a live bluegrass band to boot. The farmers simply travel a ways to get there.

2. The land. What makes ideal farmland? Well, the soil for starters, and the ideal is sandy loam, clay loam or something between. Both tend to be well draining, are easy to turn or till, hold moisture well, and are apt to have a high percentage of organic matter. Poor soils can be improved, mind you, but all other things being equal, you probably want to start with a soil that's easy to work with.

All soils contain relative amounts of sand, silt, and clay, and are classified by the feel of a handful under moist conditions. The three categories are: C (coarse textures of sand, loamy sand, and sandy loam), M (medium textures of loam and silt loam), and F (fine textures of clay loam, silty clay loam, silty clay, and clay. The finest soils I've worked are sandy loam and clay loam, but anything in between is fine.

In judging a plot of land, I'd first avoid land with little depth to bedrock. This can be determined by taking core samples, but you can also look for rock outcroppings in the surrounding area. Low-lying land is less susceptible to the problem than hillsides. But lowland can have problems with a high water table or periodic inundation from nearby rivers or even small creeks, so ask around and check with the local Soil Conservation Service or Agricultural Extension Office.

The best land would have a deep bed of topsoil, and a shovel can tell you if the land has it. Topsoil is generally darker than the lighter, underlying subsoil. The more topsoil the better, as a rule. That doesn't mean you won't need to improve the soil's productivity, but it gives you a good starting point. There are no hard, fast rules, but I prefer a good 6 to 8 inches of topsoil, if not more. And do know that over time you can increase the depth of topsoil you have.

A variety of cultivating tools for the smallholding.

Find out the property's history. If the soil has been chemically farmed, you can repair the damage over several years, but won't be able to get organic certification for three years. The ideal land is that which has been fallow or perhaps merely used for haying.

If things look good this far, get the soil tested. The tests don't cost much, but I'd get three of them and compare the results. Maybe get two from private labs, and one from the extension service. Tell each lab that you'll be using the soil for organic farming, because it may color their recommendations. Gather samples from random spots on the plot, and mix them together to be tested.

Test for pH, nutrients, and percent of organic matter. The nutrients commonly tested for are phosphorous, potassium, calcium, sulfur, and magnesium. Most labs don't test for nitrogen because its prevalence can change dramatically and quickly.

GATHER EXPERIENCE

If you're new to organics or to large-scale growing, then books won't help much, and neither will agricultural courses. Ag schools teach virtually nothing about organic growing, and while there are good books out there, there is an art, a rhythm, a deep, almost primordial sense of life that one must develop to farm successfully.

It's not the "green thumb" thing, because I'm convinced that not only do most human beings *want* to grow plants, they *can* grow them. It's a confidence thing. My old friend James Underwood Crockett, who hosted the original *Crockett's Victory Garden* show on PBS used to say, "You

When it comes to hand tools, buy the best. Your back, and your checkbook, will thank you.

Avoid Debt

America's household debt load has, for the first time, grown to where we owe more money than we make. Debt levels surpassed household income by more than 8 percent, in 2005, according to a May 2006 study by the Center for American Progress (CAP). The middle class in particular is struggling with stagnant wages and a rising cost of living.

"People are borrowing more money not because of over-consumption, but because they're caught in a bind," says Christian E. Weller, economist and author of a recent (CAP) report entitled "Drowning in Debt." Weller went on to say that "In that bind, the only escape valve for middle class families is to borrow more money."

Now I hate to take an economist to task, but it simply ricochets off my noneconomist mind how an entire nation of households can go into debt without over-consumption being greatly to blame. I'm not talking about a medical crisis or being out of work, I'm talking about *buying stuff.* Lots of stuff. Boats, houses, BlackBerries, DVDs, VCRs, iPods, you name it. You know: Stuff.

Stuff you don't need on a farm.

need six hours of sun and good soil. The rest is details."

It doesn't matter which end of the seed is down, or whether it's 1/8 inch shallower than instructed. It'll come up. It almost always does. The only seed that won't come up is the one you fail to plant.

To get this almost Zen acquaintance with growing, sign on as an intern or apprentice at a successful organic farm. Better yet, work at two farms, with two farmers, because these aren't people who agree on a lot, but both can be right, so you may as well get two solid opinions. Try a large, mechanized farm, and a small one of less than 5 acres. Learn what tools and machinery you *really* need, and what you don't need. Some farmers accumulate so many tools, tractors, and attachments over the years that you'd think they were collectors. It's not that; it's that we all keep hoping the next thing we buy (or rather, invest in) will change everything and make our farm life easier. It doesn't, but if it's any improvement at all, we'll use it and put the other aside.

You can also gain experience by simply starting a garden and making it bigger every year. That's what we did. Buy a few books. Order some seed. Scrounge around for good sources of manure, compost fodder, and mulch. Practice, practice, practice. Try new things. Talk to people. See what works and what doesn't. You'll find that one of the best feelings in the world is growing your own food. The other best feeling is selling it. When you get a surplus, put out a pair of sawhorses with a plywood top by the road. Price your produce, put out a laminated sign saying you're on the honor system, and put out a can with a few dollar bills and change. See what happens.

KEEP A SMALL NUT

By that I mean simplify. Avoid buying new cars. Avoid even thinking about a big house you can't afford. Forget restaurants; cook in. Learn how to freeze and can your produce. Avoid all gadgets like electric can openers, leaf blowers, edgers, and the like. Have a small lawn and a small lawn mower. You don't need a riding mower, you need a tiller. Till up all that extra lawn and grow something beautiful on it.

Avoid anything related to stress, like perhaps your job. Being an organic grower is the biggest de-stressor I've ever had, so why clutter things up with the opposite?

A Pulitzer-Prize–winning reporter for the *Washington Post* named Ward Sinclair

Planting highbush blueberries in early spring— an excellent market crop.

used to sell his extra organic produce to fellow workers in the office. His list of customers kept growing, until he decided that selling food was more satisfying than selling his writing, so he quit and set up what became one of the most successful market farms in Pennsylvania.

If you can, simplify your life down to the essentials.

KEEP YOUR DAY JOB

In going into organic growing, or any new venture for the first time, it helps to have at least one household income until things get up and running. This takes some of the pressure off a system that doesn't yield to pressure. Your land will most likely take at least several years to get in shape for good cropping, not to mention capital outlay for tools, machinery, equipment, and the endless other supplies needed for even a small farm. You don't want to try and rush this because it won't be rushed. Take plenty of time before you or your spouse quits a job to do the farm business plan, look at competition, plan your marketing strategy, and envision a worst-case scenario for the first five years. That's typically the time it takes for any new business to turn the corner into profitability.

We started slowly over the years, building soil and building knowledge before going into farming full-time. And then the timing (losing my job) wasn't everything it could be, but every day we invested in the front end paid off later.

Think long and hard about the long and hard hours you'll be working, and what that means in terms of dinners out and the like. If the sun's up, you're up, and when it's down you're still up. I can remember many a night picking corn by truck headlights the night before market day. If that part works for you, and the numbers look reasonable, and you have a nodding acquaintance with plants and soils, then go ahead and don't look back.

What this world needs is more organic farmers. I can't think of a more difficult and wonderful job to have.

Chapter 3

THE SOUL OF A SOIL

Peppers under plastic. NOP rules require that plastic mulch be disposed of every year lest it degrade in the soil.

There is no free lunch in nature. To grow a larger rose you lose fragrance. To get so-called hybrid vigor you lose the ability to save seeds. By employing science to subdue mosquitoes, weeds, or most any other bothersome blip in our radar screens, we invariably create a new problem that must be addressed in turn—often at a higher cost than the original one.

Whenever we try to control something that we cannot comprehend, such as nature, there will be no end of trouble.

Dr. Lewis Thomas, former president of the Sloan-Kettering Cancer Center and author of *The Lives of a Cell* (Bantam, 1975), expressed the situation this way in a discussion on taking control of one's body:

"One ought to feel elated by the prospect of taking personal charge, calling the shots, running one's cells around like toy trains. My trouble, to be candid, is a lack of confidence. If I were to be informed tomorrow that I was in direct communication with my liver, and could now take over,

Regard your soil as the living, beathing entity it is—to be nurtured, fed, and sustainted.

I would become deeply depressed. I'd sooner be told, 30,000 feet over Denver, that the 747 jet in which I had a coach seat was now mine to operate as I pleased."

And that's how it is with farming. Nature for all millennia has been nurturing her planet with organic matter, biodiversity, and an impossibly complex world of microbial activity that I, for one, haven't the hubris to challenge. Science is wondrous, mind you, but when it comes to my food, my children, and my Earth, I'll go with the original architect of my soil.

In coming to know the soul of organic growing, one must be able to think of the soil as the living, breathing entity it is. Regard the soil as you might regard your own life, or a child's—to be nurtured, fed, cared for, and sustained—and the rest pretty much takes care of itself. Oh, farming gets dicey, mind you, constantly dicey; but that's the farming, not the organics. Listen, observe, take notes, learn, have faith, and it keeps getting easier.

The next thing to understand is patience. And we're not talking sandlot patience here; we're talking Big Mama patience. We're talking p a t i e n c e. And that's the hard part because we've become a people in love with instant gratification. We want our wasp and hornet killer to knock those babies *out of the sky, dude*. We want medication to take effect now, if not sooner. We need fast food passed through the driver's side window without having to use the brakes.

What we do *not* want, at all, ever, in any way, is to have to heed Mother Nature's position on time—because Mother Nature's position is that your worn-out soil might need a long runway before it gets off the ground.

It takes time to restore conventional farmland and that's that. But it can always be done. All that's required is organic matter and time. How much of each depends on your soil, but the goal every time is to feed what's often called the soil *microherd*. By this I mean the billions of beneficial microorganisms under every square foot of healthy soil. They serve first to break down organic matter into a form plants can use, and then also improve the soil's structure and the ability of plants to draw sustenance from it. And that's only the beginning.

My old friend Eliot Coleman, in his classic work, *The New Organic Grower* (Chelsea Green, 1989) recommends five amendments for organic soil:

- **Organic matter**
- **Rock Phosphate** (a finely ground rock powder, widely available)
- **Greensand** (a material from seabed deposits, also widely available)
- **Limestone** (the rock powder used to sweeten acid soils)
- **Specific micronutrients** that may be missing from your soil, but are essential for crops

Eliot's advice is solid, and it's a rare farmer who would follow it and not do well. My own experience is that only one soil amendment from his list is truly needed,

In coming to know the soul of organic growing, one must be able to think of the soil as the living, breathing entity it is.

Hank and Cindy Delvin (shown here loading produce) were awash in chemicals 20 years ago. Now they run one of the most productive small farms in Tennessee, certified organic.

One is stable, or inert, consisting of organic matter that has decayed about as much as it can. This end-product organic matter is often called humus (a word with many definitions). The material may be a century or more old, and it plays a major role in improving soil structure and its moisture-holding capacity. Humus can absorb more than six times its weight in water, and won't give it up easily except to plant roots. In droughty sandy soil, this can spell the difference between crop failure and a good harvest. One study found that humus can hold five times the amount of nutrients than clay alone can hold.

The other form of soil organic matter is anything *but* inert. It's humming with the activity of life in the process of decay. This includes living organisms, fresh organic residue, and organic matter that is still decomposing. It is known as the active fraction of soil organic matter, and it—along with the microbes that feed on it—are vital to plant nutrition. It stimulates the growth and reproduction of beneficial microbial organisms that in turn convert this active matter into plant-available nutrients. It's a circle. These organisms also degrade pesticides and pollutants, help control disease, and—in the case of mycorrhizal fungi—help bind soil particles into larger aggregates. The resulting crumbly soil allows better root development, better moisture penetration, and of course makes tilling easier. (Mycorrhizal fungi, with their little tentacle-like hyphae, also extend greatly a plant root's ability to reach moisture and nutrients in a drought.)

A soil organic matter content of 5 percent in the top 12 inches of soil is considered a good goal for agricultural uses. Getting there only requires that you add more organic matter to soil than is lost through decomposition or erosion. Two factors most influence this:

Inputs

Organic matter can be built up in the soil through the addition of compost, organic mulch, surface residue, and from turning under cover crops. Material is also added from the decomposing root systems of plants (including weeds) that die before being turned under, or that lie below cultivation depth.

Anything you can do to boost this process, such as irrigating to promote plant growth and leaving no field unsown for long, is a plus. Organic fertilizers also boost plant growth, of

and that is organic matter—particularly if it comes from a wide array of sources. The only thing I add to our soil besides organic matter is greensand and fish/seaweed emulsion—but this is more of a safety net than anything. Our crops grow beautifully and I hate soil tests, don't have time for them, so these last two amendments assure me that we have micronutrients aplenty.

The addition of organic matter to a soil automatically shifts that soil to a healthy pH. It doesn't happen overnight, however, so if your soil pH is way out of whack, add lime until the organic matter builds up and works its magic. If you're not sure what nutrients your organic matter is providing, or missing, your crops should tell you. Or have the soil tested, or go the greensand route. It's wonderful stuff.

SOIL ORGANIC MATTERS

So what is the magic of organic matter? What is it with this stuff and why does it make us happy?

In a nutshell, organic matter is the vast array of carbon compounds in the soil, left behind by decomposing plants and other organisms. They most often come in two forms:

According to USDA statistics, the number one fresh vegetable (on a pounds-per-person basis) is lettuce.

Land that's been conventionally farmed cannot be certified organic for three years. During that time, it's considered to be in transition. But while chemical detoxification can take three years, that may be just the beginning in terms of fertility. It's impossible to say with any certainly because of all the variables.

One university extension service report on the subject stated this:

"Why does it take so long for organic matter levels to increase? An acre of soil 6 inches deep weighs about 100 tons. So increasing the proportion of organic matter from 2 to 3 percent is actually a 10-ton change. (But) you cannot simply add 10 tons of manure or residue and expect to measure a 1 percent increase in soil organic matter. Only (part) of the original matter becomes soil organic matter. Much of the rest is converted over several years to carbon dioxide."

I took this as a doomsday message: Where, I might ask, will I find 10 tons of compost to add to every farm acre annually, beginning now and ending whenever, with most of it floating away besides?

The answer of course is that soil weighs a good deal more per cubic foot than finished compost, so the comparisons fall flat.

Besides, if I don't have 10 tons of compost to add every year, then whatever I have will suffice.

It's a rule of farming, and of life.

course, and stimulate microbial activity, although I'm not sure how much actual organic matter they contribute. Doubtless this varies.

Tillage

Any time soil is turned over, the added oxygenation results in an explosion of soil microbial activity and a rapid decomposition of organic matter. This is especially true of young organic matter in the active state of decomposition, but it can also affect the stable compounds when soil aggregates are broken up by mechanical tillage. This drains your bank of organic matter.

In a nutshell, to build up organic matter in your soil, avoid cultivation—especially deep tillage—as much as possible, while feeding your land as much organic matter as you can. (There are trade-offs, of course, because deep tillage is often needed to aerate highly compacted soil.) In truly sustainable agriculture, nearly all of these inputs will come directly from the farm, including kitchen swill. As a practical matter, though, depleted soil will need off-farm inputs for a number of years to build up a rich, sustainable fertility.

Ashes to ashes: Rotted or defective produce never goes to waste in a good organic operation.

Chapter 4

COVER CROPS & GREEN MANURES

This second cutting of hay behind our house will be used for round bales.

Cover crops and green manures lie at the heart of sustainable agriculture. They have been around since the dawn of agriculture and they can work wonders. They add inexpensive soil nitrogen, boost soil organic matter, suppress weeds, improve soil structure and microbial activity, prevent erosion, capture nutrients, and help break up and bring nutrients to compact subsoil. Several studies have shown that they even attract and harbor beneficial insects.

Green manure is any cover crop that is turned under in its green state, or soon after flowering, to add nutrients to the soil. A cover crop often is considered one designed to cover bare soil—whether it's mowed, turned under, or left to die back. The words are often used interchangeably however.

This field, already harvested and disked in midsummer, will be planted with a cover crop to smother weeds, prevent erosion, and add more organic matter to the soil.

Traditionally, cover crops were sown in the fall and tilled under, rolled, or hayed in the spring before main-season crops were planted. Winter rye is the standard winter cover crop for cold areas, and is relatively inexpensive. High-value legumes such as vetch can cost three times as much. But the types and uses of cover crops have increased dramatically in recent years, presumably with the increase in organics.

In selecting which cover crops and green manures to sow, look first at what you're trying to accomplish.

Let's start with nitrogen.

NITROGEN

The roots of legumes take nitrogen out of air in the soil and store it in their plant tissues.

Leguminous green-manure crops can add anywhere from 40 to 200 pounds of nitrogen to an acre of soil, depending on planting time, soil moisture, pH, and other factors. About half of this will be released when the legume is turned under, and about half will be available to a following grain or vegetable crop. Enough residual nitrogen will remain in the soil to boost yields for two or three years.

Forage legumes serve double duty in a rotation from their use as hay or pasture grasses, while still providing nitrogen from their root mass and regrowth. Alfalfa, followed by red clover, produces the most root mass and the least top growth, while soybeans and vetch produce the most tops and least root matter.

Crops with the highest nitrogen-fixing capacity are alfalfa, hairy vetch, and cowpeas. Medium producers are crimson clover, field peas, white clover, and red clover. Beans have the least nitrogen-fixing capacity among legumes. Soybeans make an

Crops with the highest nitrogen-fixing capacity are alfalfa, hairy vetch, and cowpeas.

The grass between this wide row of mulched beans may not be attractive, but it's not hurting the beans. I know of no organic grower whose farm would win a beauty contest, and yet they are things of beauty.

excellent undersown crop with corn, depending on how tight your corn is planted, because they can stand some shade and help suppress weeds.

These legumes have a minimal long-range effect on soil organic matter levels.

SOIL ORGANIC MATTER

Cover crops can add to an acre of soil the equivalent of 10 tons of fresh manure or about 2 tons of dry manure matter, according to 1984 tests by the Woods End Agricultural Institute in Maine. What's more, this nitrogen is right in the soil where you want it, not blowing off to Kansas.

The crops grown to increase soil organic matter are those with a relatively large above-ground mass. They include annual

ryegrass, cereal rye, triticale, sorghum/Sudan grass, and hairy vetch. Grasses contain a substance called lignin, which makes them more resistant to break-down than legumes, so they stay longer in the soil as organic matter. These grasses not only add organic matter to the soil, but as they break down they exude substances that help bind together soil particles as granules, or aggregates. This makes the soil easier to work and more able to absorb and store water. It also allows better soil aeration and reduces soil surface crusting.

It will take several years for grass cover crops to significantly build up soil organic matter levels, because much depends on climate, existing soil structure, and other factors. But you have to start somewhere, and every bit helps.

MICROBIAL ACTIVITY

When a succulent young green manure crop is turned under, soil microbes multiply rapidly, breaking down the new plant material, releasing nutrients that can be used by subsequent crops. How quickly this happens depends on soil moisture, temperature, and the nature of the plant material. Warm temperatures increase microbial activity, and overly dry or wet conditions hinder it.

The nature of the plant material has to do with its carbon/nitrogen ratio (see sidebar). Organic matter decomposes fastest when it's between 15:1 and 25:1 carbon/nitrogen ratio. Hairy vetch can have an extremely low C:N ratio of 10:1, whereas dry corn stalks are about 60:1. These last will take a good while to decompose unless composted in the company of green matter, such as fresh vetch.

NUTRIENT RELEASE

Cover crops also capture and make available nutrients other than nitrogen. These include phosphorous (P), potassium (K), calcium (Ca), magnesium (Mg), sulfur (S), and other nutrients that the cover crop scavenges from the soil and makes available as it decomposes. A crop with a large, deep root system that grows quickly helps prevent nutrients from leaching away after a cash crop is harvested. Annual ryegrass, triticale, and alfalfa are good choices. Deep-root cover crops also bring up nutrients from the subsoil, making them more available in the top 12 inches of soil surface to succeeding crops. Buckwheat, lupine, and

One-third of the raw materials used for all purposes in the United States goes to the production of meats, dairy products, and eggs.

sweet clover are noted for their ability to extract phosphorous from soils, according to "Overview of Cover Crops and Green Manures," from the National Sustainable Agriculture Information Service.

WEED CONTROL

A wide variety of cover crops provide weed control while adding organic matter to the soil. These suppress weeds by shading and competition. When killed by rolling or other means, they continue to suppress weeds by blocking out light.

Some cover crops suppress weeds chemically by relaxing compounds that help prevent the germination or growth of weed seeds. This toxicity is known as allelopathy.

Among the grasses found or reported to have allelopathic properties are buckwheat, wheat, barley, rye, hairy vetch, red clover, sorghum/Sudan grass and members of the mustard family.

Some cover crops, such as dwarf white clover or vetch, interplanted between rows of a main-season crop, not only help suppress weeds and reduce the need for cultivation, but help keep soil from spattering on your eggplants or lettuce during a rain.

A cover crop is planted and mowed between rows of strawberries. Black plastic mulch is used in this case to collect solar heat. *Courtesy of Delvin Farms*

Carbon/Nitrogen Ratio

The carbon/nitrogen ratio of organic matter is crucial to knowing how fast it will decompose or make its nitrogen available to crops. High carbon material is generally mature, fibrous, and dry material such as straw, leaves, peat, and wood chips. Nitrogen-rich materials are the "greens—grass clippings, kitchen wastes, and succulent green cover crops.

The thing to know is that the higher the C:N ratio, the longer it will take to decompose and release its available nitrogen. Ratios higher than 30:1 will tie up nitrogen. Those lower than 20:1 will release nitrogen. A ratio of 25:1 is considered stable.

It's a good ratio to aim for in making compost.

The ratio for fresh sawdust and wood chips is more than 300:1. This stuff won't be fertilizing your crops for a long time. I am just now using compost that started out as a massive pile of wood chips eight years ago.

The C:N ratio for legumes ranges from about 9:1 to 19:1, meaning their nitrogen will be available to other plants quickly. The C:N ratio for peat moss is 45:1 and for straw it can range from 20:1 to about 50:1. Dry corn stalks weigh in at 60:1.

CROP ROTATION

Chapter 5

While vegetables, grains, and other crops can and should be rotated, perennial, vegetables, berries (such as these thornless blackberries) and orchard trees cannot. The soil here will need special attention.

Crop rotation, like green manures, dates back to the dawn of agriculture. It fell into disuse with the advent of modern chemical fertilizers and monocropping, but this was to the peril of soil health, so rotational cropping is considered a vital spoke in the wheel of organic farming today.

In its simplest form, crop rotation is the act of planting something different in a field every year. A rotational sequence is

usually for a series of years, coming back to the original crop after four or more years. This cycle disrupts weed and pest activity, helps eliminate soil-borne diseases, and aids soil fertility by growing crops with different nutrient needs in a field (thereby not using up one nutrient such as nitrogen by repeatedly planting corn in the same plot). Some crops do well if rotated after other crops, and when

The first key to a good rotational plan is to vary the crop family on any one field or area.

green manures and cover crops are included in the rotation, rotational planting provides the best of all worlds.

Eliot Coleman states flatly that crop rotation is "the single most important practice in a multi-cropping program."

The idea is not just to vary the *crops* in a field, but the family of crops. Rotating cabbage after broccoli does no good, for example, because both are in the brassica family, subject to the same pests, diseases, and nutrient needs.

So the first key to a good rotational plan is to vary the crop family on any one field or area.

The simplest rotation would be an eight-year cycle, with the major plant families occupying one section every year. This may not be practical, however, because many areas may be sown to succession crops, and a crop such as corn takes up considerably more space than lettuce.

It's also wise to plant corn in the year following a leguminous vegetable or green manure crop, or a brassica undersown with clover. Most any crop is thought to benefit by following onions or leeks. Root crops such as leeks and carrots are good to sow following a crop of potatoes or winter squash, because the former are greatly

Vegetable Families

Chenopodiaceae
Beet, Chard, Spinach

Solanaceae
Potato, Tomato, Pepper

Cucurbitaceae
Cucumber, melon, squash

Alliaceae
Garlic, Onion, Leek, Shallot

Leguminosae (Fabaceae)
Alfalfa, Beans, Peas, Clover,
Vetch, Lentils

Umbelliferae
Carrot, Celeriac, Celery,
Fennel, Parsley, Parsnip, Dill

Compositae (Asteraceae)
Endive, Jerusalem Artichoke,
Lettuce, Salsify

Brassicaceae (Cruciferae)
Broccoli, Cabbage, Cauliflower, Kale,
Radish, Turnip, Brussels Sprouts,
Oriental brassicas, Mustard, Kohlrabi

Miscellaneous
Corn, New Zealand Spinach, Rye,
Buckwheat, Mâche

Parsley is commonly used as a garnish, but it also makes a nice breath freshener and is arguably the most nutritious plant you'll find.

It's always wise to offer variety, but too much variety makes customers shy away.

Peppers under white plastic. Note the use of rot-resistant wood stakes rather than metal T-posts.

affected by weeds, and the latter generally smother weeds (in preparation for slow-growing root vegetables later in the succession). If squash follows potatoes in the lineup, then you get two almost-weed-free years before leeks and carrots. Or, a good weed-suppressing cover crop may be planted prior to weed-prone crops.

Factor in green manures and things can really start to get wild. Figuring out a rotation is worth every winter moment you can devote to it because of the benefits. And all of this calculating is free. It's like free compost or seed in what it can do for your crops.

An eight-year rotation, with one crop on one of eight fields each year, yields 5,040 variations that can be tried. Don't be

alarmed. The key is simply to make your crops a moving target and keep every field sown to green manures and cover crops as often in the rotation as possible.

For example, in the months following a spring crop harvest, and the summer sowing of a fall crop, you might well find time to sow a quick-growing warm-season cover crop such as buckwheat or cowpeas.

Many vegetable crops can be overseeded with a cover crop, which will then be established and growing after vegetable harvest. Select cover crops that tolerate some shade and harvest traffic, such as dwarf white clover or vetch, especially where there will be multiple pickings, such as in tomatoes or peppers. Brassicas, which get little foot traffic, may be overseeded with other choices.

The average produce item in the United States travels 1,200 miles from farm to market.

Chapter 6

THE COMPOST QUESTION

Mulching beds in the spring. Spreading compost on a spring field prior to seeding. *Courtesy of Delvin Farms*

Composting on a backyard level is easy: Toss dead vegetation on the pile and let it rot. I like to keep things simple.

On the farm level, composting is a little different, and some farmers don't want to mess with it. For one thing, you need a lot more material. You also need a way to get it to the farm, a place to put it, and a way to turn it. Oh, and a way to spread it. I do this either with a yard cart, or by bucket-loading it into the pickup and hand-spreading in the field—a small manure spreader would be very helpful to the process.

When we started composting on a large scale, I let word get out to local landscape contractors that I wanted their leaves and untreated grass clippings. This was a win-win for them because the town recycling center charged them to dump it, and I was free.

I didn't know what to expect, however, and quickly learned that I had two choices: continue farming, or become a compost company. By the time leaves covered more than half an acre in piles often 6 feet high, I called time out. Fortunately, the leaves and grass came in at almost the precise 25:1 carbon/nitrogen ratio I wanted. The manure from our own horses and chickens, along with cow manure from a neighboring dairy farm, rounded out the mixture.

I decided to try to hot-compost the piles, so I began mixing and turning the material with a tractor bucket in the path of two large sprinklers, but the system was unwieldy, so I simply mixed it all and let it rot. The piles were ready in about a year.

We've also sought out other sources of compost materials, including nearby restaurants for kitchen swill and Starbucks for coffee grounds. This may or may not be time well spent, but it's habit.

Doing this today, however, under the National Organic Program (NOP) for organic farm certification, the rules are a bit different, as regards manure in compost.

The NOP has put no specific restrictions on when farmers can apply composted manure to crops, but it is quite specific about manure-composting procedures. According to the NOP regulations, compost must meet the following criteria:

An initial carbon/nitrogen ratio of between 25:1 and 40:1 must exist for the blend of materials in the "pile."

Temperatures between 131 degrees F and 170 degrees F must be sustained for 3 days using an in-vessel or static-aerated pile system; or temperatures between 131 degrees F and 170 degrees F must be sustained for 15 days using a windrow composting system, during which period the materials must be turned a minimum of five times.

Now *that's* specific.

These hot-composting procedures will kill any pathogens and most weed seeds. Any composted manure that is not composted as discussed above must be considered raw manure, in which case it cannot be applied to a field within 120 days of harvest for a crop where the edible portion touches the soil (such as carrots), or within 90 days of harvest where the edible portion does not touch the soil (such as corn).

The NOP goes on to say, "Composting livestock manure reduces many of the drawbacks associated with raw manure use. Good compost is a "safe" fertilizer. Low in soluble salts, it doesn't "burn" plants. It's also less likely to cause nutrient imbalances. It can safely be applied directly to growing vegetable crops. Many commercially available organic fertilizers are based on composted animal manures

King Compost. Most any organic material may be added to the pile, but carbon and nitrogen must be in balance for the most efficient use of the materials. NOP rules strictly govern manure composting. This composting pile, comprised largely of pine chip bedding and livestock manure, must be treated differently than compost that contains no manure, under the organic certification rules.

This compost (near the interstate) has been aging for months, waiting to be applied next spring.

Carbon/Nitrogen Ratios

Material	%C	%N	C/N
Spruce sawdust	50	0.05	600/1
Wheat straw	38	0.5	80/1
Corn stover	40	0.7	57/1
Rye cover crop, anthesis	40	1.1	37/1
Bluegrass—lawn	40	1.3	31/1
Rye cover crop, vegetative	40	1.5	26/1
Mature alfalfa hay	40	1.8	25/1
Rotted barnyard manure	41	2.1	20/1
Young alfalfa hay	40	3.0	13/1
Hairy vetch cover crop	40	3.5	11/1
Digested sludge	31	4.5	7/1
Soil organic matter	56	4.9	11/1
Grassland	*%C*	*%N*	*C/N*
Forest	52	2.3	23/1
Subsoil	46	5.1	9/1
Forest soils	3.6	0.18	20/1
Prairie/grass	2.8	0.22	13/1

supplemented with rock powders, plant byproducts like alfalfa meal, and additional animal byproducts like blood, bone, and feather meals."

It warns, however, that composted broiler (chicken) litter—though more stable than raw litter—will be abundant in phosphates and low in calcium. Continued applications may lead to imbalanced soil conditions in the long term, as with some raw manures. The NOP strongly advises compost testing to monitor nutrient levels.

A more recent NOP concern is the inclusion of additional copper in poultry diets and its accumulation in the excreted manure. Copper is an essential plant nutrient, but an excessive level in the soil is toxic. This concern is most relevant to organic horticultural producers who often apply significant amounts of copper as fungicides and bactericides, increasing the hazard of buildup in the soil. Whenever you import large amounts of either composted

Relative Values of Compost Materials

Material	Nitrogen (N) %	Phosphrous (P) %	Potassium (K) %	Humus potential	Breakdown rate	Comments
Natural Manures	*Note: Manures vary widely, depending on bedding or litter content. Some other materials also vary.*					
Cattle	0.4	0.3	0.44	high	moderate	Can be weedy. Composts well.
Horse	0.7	0.5	0.6	high	moderate	Best on clay soils. Hot fresh.
Sheep, goats	0.65	0.45	0.25	medium high	moderate	Hot when fresh.
Pig	0.35	0.45	0.45	medium low	moderate	
Sheep, goats	0.65	0.45	0.25	medium high	moderate	Hot when fresh.
Duck	1.1	1.45	0.5	medium low	moderate	Hot when fresh.
Goose	1.1	0.5	0.5	medium low	moderate	Hot when fresh.
Chicken	1.5	1	0.5	medium low	moderate	Hot when fresh.
Turkey	1.3	0.7	0.5	medium low	moderate	Hot when fresh.
Rabbit	2	1	0.5	medium low	moderate	Hot when fresh.
Processed manures						
Cattle	1.3	0.9	0.8	medium	moderate	Generally odorless, weed free.
Sheep, goat	2	1.3	2	medium	moderate	Generally odorless, weed free.
Poultry	4.5	3.2	1.3	low	moderate	Weed free
Livestock and poultry bedding						
Oat straw	0.06	0.2	1.3	medium high	moderate	All straws can contain weed seeds.
Rye straw	0.6	0.3	0.8	medium high	moderate	
Wheat straw	0.5	0.2	1	medium high	moderate	
Sawdust	0.2	0.1	0.2	medium	moderate fast	Wood products can reduce available soil N
Wod shavings	0.2	0.1	0.2	medium	moderate slow	See above
Tree leaves	0.8	0.35	0.2	medium	moderate slow	Black walnut leaves contain plant toxins.
Pine needles	0.45	0.12	0.3	low	moderate slow	Tends to be acidic
Sugar cane	1	0.1	1.3	high	moderate slow	
Peat moss	1	trace	trace	low	moderate slow	Tends to be acidic
Other vegetable sources						
Alfalfa meal	2.5	0.5	2	high	moderate fast	
Alfalfa straw	1.5	0.3	1.5	high	moderate	
Coca shells	2.5	1	3	medium	moderate	
Cottonseed meal	7.2	2.5	1.7	high	moderate fast	Acidic; may contain pescicide residues
Linseed meal	5.5	1.7	1.3	high	moderate fast	
Soybean meal	7	2	2	high	rapid	
winery pomace	1.5	1.5	0.75	medium	moderate fast	
Coffee grounds	2.1	0.3	0.3	moderate low	moderate	Starbucks will give it away.
Tea leaves	4.2	0.6	0.4	moderate high	moderate fast	The bags also compost well.
Cornstalks	0.5	0.3	1	moderate	very slow	Shred them for faster breakdown
Peanut shells	1	0.2	0.75	medium	moderate	Chop them up.
Seaweed	0.5	0.1	2.5	moderate	slow	Great source of micronutrients
Sugar beet leaves	0.3	0.4	trace	medium	moderate fast	
Tobacco waste	2.2	0.5	6	moderate high	moderate slow	
Hay						
Salt marsh hay	1.1	0.3	0.75	low	very slow	Free of fresh-water weed seeds.
Timothy etc.	1.3	0.6	2	moderate high	moderate fast	May contain numerous weed seeds.
Non-vegetable sources						
Ashes, hardwood	0	1.5	7	none	ready	You'll need little, if any
Dried blood	9.0-14.0	1	1	none	ready	
Bone meal	3.5	22	0	none	very slow	I don't think it's worth the money.
Egg shells	1.2	0.4	0.1	none	slow	
Feathers	12.5	0	0	low	moderate	
Fish emulsion	5	2.4	1.2	minor	moderate	Rich in micronutrients.
Fish scrap	9	6	2.5	low	moderate	Attracts animals.
Hair	9	0	0	low	slow	

Green manures and cover crops

These are generally not used as compost fodder. Green manure crops are turned under while green or just after flowering to improve soil structure, add organic matter, or suppress weeds. Legume green manures such as sweet clover and soybeansalso add nitrogen to the soil. Cover crops suppress weeds, prevent erosion, and reduce insect pests and diseases -- and may be turned under for soil improvement.

or raw manure onto the farm, it is wise to inquire about the feeding practices at the source, or have the material tested.

NOP further cautions that "while composting can degrade many organic contaminants, it cannot eliminate heavy metals. In fact, composting concentrates metals, making the contaminated compost, pound for pound, more potentially hazardous than the manure it was created from. Broiler litter and broiler litter composts have been restricted from certified organic production in the mid-South largely for this reason. Arsenic—once used in chicken feed as an appetite stimulant and antibiotic—was a particular concern. Since the precise composition of commercial livestock feeds is proprietary information, arsenic may still be an additive in formulations in some regions."

In a word, know what's going into your compost, and use as many on-farm resources as possible. Leaves are always safe—so far as I can determine—as are untreated grass clippings. My only hesitation with the latter is that grass clippings often arrive soaking wet, and if they aren't spread to dry immediately, anaerobic bacteria will explode inside the pile and make it smelly and almost worthless.

If you get grass delivered, you've got to spread it that day, not the next. Then add it to your compost.

I know it's worth it, because the composting process turns raw products into humus—the relatively stable, nutrient-rich, and chemically active organic fraction found in good organic soil. The more you have of this, the healthier your plants will be. In this stable humus, in NOP's words, "there is practically no free ammonia or soluble nitrate, but a large amount of nitrogen is tied up as proteins, amino acids, and other biological components. Other nutrients are stabilized in compost as well."

It's like having a savings account. All materials listed on the previous page benefit soil fertility to one extent or another. "Humus" here is defined as organic matter that is fully broken down and adds immeasurably to soil structure, or tilth.

Autumn leaves, considered a bother by many, are a key ingredient in our mulching and composting system.

Worm Tea: The Magic Elixer

Worm castings have been long known to be a highly fertile component of organic soil. Compared to average soil, worm castings are said to contain 5 times more nitrogen, 7 times more phosphorus and 11 times more potassium. They are rich in humic acids and improve the structure of the soil as well.

Many organic growers buy castings as an off-farm input, while others obtain the material by operating a worm composting (or vermicomposting) system in their home or heated outbuilding. The system amounts to feeding worms your kitchen scraps in an enclosed bin, and harvesting their castings for use as organic fertilizer.

Unfortunately the castings are rather difficult to harvest without also harvesting baby worms and worm eggs. Nor could it be called a tidy enterprise. What's more, vermi-composting bins available from mail order sources can cost $100 or more, and that's for a small-scale backyard garden operation.

We get around these limitations by harvesting, not the castings, but what I call the "worm tea" that results from a vermicomposting system. We use this as a foliar feed, and the results have been remarkable. Here is how our system works:

Step 1
Start with several 10-gallon plastic storage tubs with lids (they cost around $5 each). Drill a dozen 1/2-inch holes in the lid, then cover it with window screening held in place with adhesive caulk. This provides aeration without allowing flies in. On the bottom of the tub, at one end, drill another 10 or so holes for drainage. Covered this with screening to keep the worms from getting out. Suspend the tubs on a shelf about 16 inches off the ground, held up with concrete blocks.

Step 2
Drill a 5/8-inch hole in the top of a gallon milk container. Into this, slide a piece of 5/8-inch (outside diameter) hollow tubing. A plastic funnel fits into the top end of the tube, and this whole thing should be set under the tub's drain holes.

Step 3
Now fill each tub about 3/4 full with rotting straw, partially rotted leaves, hay, shredded cardboard, or even shredded newspaper. Wet the material down well until it's fully damp, then place in each tub 500 or so red wriggler worms (Eisenia foetida), which are ideally suited to life in a worm bin. They can digest large amounts of organic matter, reproduce rapidly, and are tolerant of different growing conditions (but should be kept at temperatures above 32 degrees Fahrenheit, and less and 90. The worms are available for mail order through any number of companies found on the Internet. We paid about $20 per 500 worms, plus shipping.

Step 4
Once the worms are in place, start adding kitchen scraps, placed under a layer of bedding, not on top. Virtually any fruit or vegetable scrap is fine, along with eggshells and coffee grounds. Avoid meat and fatty foods. Start with small amounts of scraps, then increase the feeding as the worms grow and multiply. There are no hard rules; use your judgment. And keep the upper layer of straw moist, dampening it every few days.

Step 5
Within a few weeks, dark brown worm juice begins to fill the empty milk jugs. I used this as a transplant solution at first in our organic vegetable beds and it seemed to work quite well, diluted by about a 1/2 cup per gallon of water. Then I tested it as a foliar feed on a row of beans. Within two days, the treated beans were at least twice the size of the non-treated plants. I tested it again as a foliar feed for cole crop and lettuce seedlings; same story. There is something amazing to this stuff.

It basically amounts to worm casting tea, and is far easier to collect and use than the castings themselves, as mentioned above. Every few weeks I simply harvest the juice and let the milk jugs fill again.

Step 6
Within a few months, you should be able to take half the bedding and worms out of a tub and start a new tub. Worms don't like their own castings, so they'll begin to die off if you don't transfer some. They don't like ordinary soil unless it's extremely rich in organic matter, so if you don't start a new worm bin, then put the overflow in a compost pile or leaf pile where they can keep working their magic.
A new worm bin and parts costs only about $6, and considering what you get, I'd expand as much as you can, based on how much room and kitchen scraps you have.
Janet wants to start packaging and selling the stuff, but I don't know. Best to just use it on our seedlings and transplants I say.

Chapter 7

GREENHOUSES

Cole crops and lettuce are started in summer for late season crops. The winter squash on the floor was a volunteer.

One of the first lessons I learned as a market grower is to raise everything you grow from seed. Everything. We buy nothing as transplants, because the first key to a strong plant is a healthy seedling—and if I haven't fed and tended that seedling, then I don't know much about it.

Moreover, seed is cheaper than plants, available in more varieties, can be certified organic, and almost certainly will contain none of the diseases and pests that so often infest commercial transplants.

Moreover still, the transplants you find at garden centers or commercial greenhouses are raised to look good on the shelf, not to succeed as a crop. The transplants we raised organically in our own greenhouses, on the other hand, quickly developed a reputation for success. Tomato and pepper lovers especially came back every year for

This is one way to fill your flats quickly with potting soil. I prefer working with a large grain scoop. The wetting agents in most commercial potting soils are not permitted in a certified organic operation.

more because our plants bore abundant fruit compared to store-bought transplants.

In direct marketing, a reputation is everything and it's a lot cheaper than advertising.

But whether or not you're considering a small nursery business, do consider a greenhouse. My own favorite heated greenhouse is a 48-foot hoop house covered in four-year poly (it lasts at least four years), with a heat source and ventilation. I can grow 10,000 seedlings in here in 32-cell trays, or more if I use 48- or 64-cell flats. Depending on your winter climate, the heat can be costly, but can be a good investment in getting to your local market early with long-season crops such as tomatoes and peppers.

We also get a good jump on the season with cole crops (any member of the crucifer family, such as cabbage, kale, or broccoli), along with all manner of herbs and cut flowers. Lettuce and other greens, corn,

peas, beans, other legumes, most vine vegetables, and all root crops don't do as well as transplants—either because they don't transplant well (most root crops and vine veggies) or are too easy to direct-seed to make transplanting worth it.

A nearby farmer had some success growing a few thousand transplant beets for our farmers' market one year, but complained loudly about the effort and hasn't done it since. Still, it can be worth experimenting. What's never worked for me might work for you. That's the joy of working in the soil. There's no wrong way to grow things.

But I'm jumping ahead. First a few basics.

PARTS AND PIECES

Most modern greenhouses are fashioned from hollow steel "bows" (a combined wall stud and rafter, in effect) spaced usually 4 to

My own favorite heated greenhouse is a 48-foot hoop house covered in four-year poly (it lasts at least four years), with a heat source and ventilation.

Only a small green house is required for raising transplants—in this case tomatoes, peppers and basil.
Courtesy Eaton's Creek Organics

6 feet apart, connected with overhead purlins or other horizontal connectors. Diagonal wind braces give the structure rigidity. Over this goes a sheet of 4- to 6-millimeter poly, which also covers the end walls. They can be virtually any width or length. A 20-by-50-foot structure will cost from $1,000 and up, with another $350 or so for the poly. Doors are up to you. Some hoop houses come with doors, but I simply buy wooden doors from a salvage yard and mount them on a 2x4 frame.

A typical hoop house propane heater, hung high on one end, with an exit fan at the other.

A typical *heated* greenhouse, on the other hand, will have a propane heater, electricity, automatic inlet vent, outlet fan, automatic temperature controls, shelves for seed flats, and a weed-impervious groundcover underfoot—the cost of these appliances and shelving range too wide to even approximate them. The structure also will typically have an outer layer of poly inflated by air from a fan to reduce heat loss. Poly (and most glass) has an R-value of 1. Two layers provide an R-value of 2. That's twice the resistance to heat loss—a savings of perhaps 30 percent.

That's how most do it, but it's not how it must be done. I've seen greenhouses made out of a simple wooden-stud frame and poly, with heat sources ranging from kerosene heaters to coal stoves. Some store the sun's heat in barrels of water to take the chill off a cold night. There is no one right way to do it.

My only advice is that propane and natural gas heaters made now are quite efficient, and automatic two-stage fans and louvers (or shutters) can keep a greenhouse full of plants from overheating and dying on a day when you can't get there to ventilate manually. If nothing else, an automatic system provides good insurance when you need it. And when you don't, you pay nothing.

That doesn't mean not to march to your own drummer and use your own ingenuity. That's what farming, and particularly organics, is all about.

TWO TRICKS

In our first hoop house I wanted to save the cost of a fan and the electricity to operate it, so I collected free from our lumberyard the flat wooden "stickers" used to separate vertical layers of lumber. Then, rather than install the second "skin" on the outside, I put it on the inside, held in place with two-sided carpet tape until I could screw the long "stickers" into place as battens. It worked like a charm.

I also discovered, in putting up shelves of cinderblock and rough-sawn pine, that the low arc of the winter sun meant I could stack shelves atop other shelves and still provide light for the first-floor plants. This increased our growing space by about 30 percent. However, as the sun's arc rises, the sunny bottom plants began to be shaded, so I had to move cold-hardy plants outside to make room.

Spring transplants in a heated greenhouse, almost ready for the field. *Courtesy of Delvin Farms*

GREENHOUSE POSITION

In locating your greenhouse, consider the following:

Soils. This is crucial if you'll be raising crops, because space is at a premium in a greenhouse. Every good square inch should be put to use, with one crop after another, and such intensive planting only works if the soil is extremely rich and friable. Packed clay won't do. So pick a site where the soil is fertile, or can be made fertile. If you'll need heavy equipment to bring in compost or other amendments, do it *before* the hoop house is up.

Sun. You want full access to the winter sun—at least six hours, and the more the

Winter is the busy season in our greenhouses, where we begin sowing perennial herbs and flowers in December for spring sales.

better. Remember that even deciduous trees can diminish solar gain in winter, and that the sun's low arc might be shaded by buildings or evergreens. In nearly all cases (the Southwest may be an exception), orient your greenhouse on an east-west axis to take full advantage of the sun's direct rays. The more intense light is that which enters directly perpendicular to a glazing, and not at an angle, where some radiation glances off. So, go east-west with the ends.

Grade. You want a flat surface (or one that can be made flat) and good drainage. If you're grading a slope for your greenhouse,

Dealing with Snow

In areas of heavy snowfall, you'll want either a heated greenhouse to melt snow and keep it from demolishing your structure, or have some provision for keeping an unheated house safe from a snow load. We use several push brooms with extensions. It's also helpful to keep a kerosene, coal, or wood heater in the greenhouse to fire up when snow or ice threaten. A half-circle hoop house is somewhat more susceptible to snow damage because the top is relatively flat, compared to the peaked, gable roof of a Gothic-arch structure.

In times of heavy snow, maneuver your heated greenhouse through the storm thusly:

If outside temperatures are below about 25 degrees Fahrenheit, and the snow is heavy (building up to more than 6 inches), turn off the inflation fan for the outer poly—especially if there's little or no wind. Otherwise snow may accumulate faster than interior heat can melt it and cause it to slide off.

The weight of accumulating snow will eventually force the outer layer into the first, mind you, but it may be too late for interior temperatures to melt away the thick snow pack.

If less than 6 inches of snow has built up, or wind is blowing the snow away, and outside temperatures are closer to freezing, then it's probably safe to leave the outer poly inflated. If snow will be falling all night, however, and you want to get any sleep, then turn the inflation off. It's better to pay for the extra heat than risk losing your greenhouse. The same is true during a heavy ice storm.

then be sure to use perforated PVC drain pipe or some other uphill system to keep water from washing away your plants in a heavy storm. A flat surface is important for several reasons, not the least of which is that if you're raising transplants in bottom-watered flats (always a good idea) and the flat isn't flat, some plants will get more water than others. We learned this the hard way.

Permits. In most cases, hoop houses are considered temporary agricultural structures, so no building permit is required. In fact, I've never heard of a permit being required. But check anyway.

Water. You'll need water, and you'll probably need it through the winter. I've heard of people running hoses into a greenhouse from a distant yard hydrant, but you won't get away with that in Zone 6 or lower and still end up cheerful on a cold day—once a hose freezes, it takes a long time to thaw. So, bring water in underground if you're going to be using the greenhouse in midwinter. This, along with rich, friable soil, is a must.

Wind. Greenhouses tend to be extremely well designed these days, but I'm not the first one who's seen 100 feet of poly blow away across a field, so everything else being equal, find a calm spot. And when installing the poly, don't leave any loose edges. Make it taut.

Spawning. Greenhouses seem to spawn when you aren't looking. First you have one, then two, then four, and so on. Keep this in mind when siting the first one. It may have company later.

THE HEATED OR TRANSPLANT HOUSE

A heated greenhouse needn't cost a fortune in fuel if you remember a few things. One, if you're raising transplants, even though a particular variety might need 70-degree temperatures to germinate, you don't (usually) have to keep it at 70 degrees all night long once true leaves appear on the seedlings. Experiment. Follow Nature. Let night temperatures drop to 55 and watch your plants do just fine. Do keep a sharp eye out for damping-off disease, however, and be sure to water only in the morning, enough that the soil surface can be dry during the cooler night hours.

If some plants need especially warm temperatures to germinate, with little variation night/day, then germinate them some-

where else and move them to the greenhouse later.

Once a plant has its first true leaves, it won't mind chilly evenings at all. The zinnia is about the only greenhouse plant that's given me any problem with damping-off disease due to cool evenings, and the problem was very minor.

Second, don't fire up the heat in the greenhouse until you absolutely have to. My experience is that petunias take about 12 weeks from seed to bloom, and most plants I grow take considerably less, so we don't heat up the greenhouse until mid-January to sell plants in April.

Remember too that it's imperative that the house be as draft-free as possible. Patch every small gap and crack you find. I've even weather-stripped the vertical edges of inlet and outlet shutters to prevent air infiltration. I also mulch heavily around the house to keep ground frost at bay.

REPOTTING AND LABELING

When growing transplants, know that it can cost an inordinate amount of time to repot even a small crop into larger pots. Therefore, always try to plant seeds in the same pot you'll be selling or transplanting from. It just makes life easier.

Always label every flat before it leaves the potting table. I know: You think you'll remember. No, you won't. Label them either before seeding, or just after. We use

Hoop house tomatoes, or field tomatoes for that matter, can be efficiently suspended with the Florida weave system shown here. *Courtesy of Eaton's Creek Organics*

Beefsteak tomatoes supported by the Florida weave system of support, with two or three plants (depending on variety) supported between two posts.

white plastic labels and waterproof pens. Include the following information on the label: plant variety, date seeded, days to germinate, and the seed source (J for Johnny's, for example, or SOC for Seeds of Change). Also keep a master list of how many flats were seeded for every plant variety, based on what is sold or used as transplants the previous year.

And finally, everybody everywhere will tell you that greenhouse pots and inserts should never be reused without first being disinfected with bleach, lest you spread diseases. They all say it, I swear. But not having time for that, I originally used to put them out in the sun to disinfect, but they'd blow everywhere and look like the dickens. Then I thought to myself, "How are my inserts going to spread disease to new plants if the old ones weren't diseased?" Duh. So now we've been reusing pots for years and never had a

problem. The only pots I absolutely *won't* use are those from a garden center or some other nursery.

Who wants to be buying new pots, flats, and inserts every year, I ask you?

THE UNHEATED GREENHOUSE

Unheated greenhouses are generally used for raising crops to harvest. That isn't always true, but it's true enough to go with for now.

Depending on your climate, you'll need either one or two layers of cold protection. An inflated outer poly cover is one option for particularly cold climates, or one can use a single poly layer, with additional frost protection from either small poly-covered tunnels inside the greenhouse, or floating row covers. Both offer only a few extra degrees of protection, but that can make a big difference.

The unheated house is, in effect, a season extender. It enables an earlier spring crop and provides protection for fall and winter harvests. It's an ideal place to start a main-season crop early enough in spring to have a harvest to market well ahead of field-grown produce. This can mean a cost premium in direct marketing through a farmers' market or roadside stand. It won't affect the bottom line of a CSA, because income is already set, but you can only imagine how grateful customers are to have that first beefsteak tomato of the year.

The unheated house is also a perfect place to raise many cool-weather crops for sale during the cold months. More on cool-weather crops later.

WARM-WEATHER PLANTS

Below are details on some of the easiest and most profitable crops to grow in an unheated house for late spring or summer sale:

Cucumbers: Train these up a trellis to save space, giving the maximum yield per square foot of greenhouse space. Many seed catalogs carry flexible trellising material for use in a hoop house or in the field.

Sow cucumbers in a heated house about four weeks before transplanting into an unheated greenhouse. We transplant about three to four weeks before the average annual last frost date. Both inside and outside we space our plants closer (12 inches or less) than is generally recommended (24 inches or more), but our soil is rich and the cukes do fine. Rows should be 5 feet or so apart.

In the greenhouse, the trellis or twine can be fastened to an overhead purlin. If a purlin isn't where you need it, install one with self-tapping screws and strapping or 2x4s. Cukes need plenty of water, and must be picked daily once they come in. I plant all sorts of varieties, as we use many pickling types for preserves, but the best salad cukes are the thin- or tender-skinned varieties.

Cut flowers: These can be very successfully grown in a greenhouse, and make a real hit at the farm stand or farmers' market. They can be show-stoppers, and help get you noticed. My favorites are delphiniums, snapdragons, cosmos, zinnias, stock, and sunflowers, but there are plenty more. Grow only annuals to allow for crop rotation, and select smaller cut-flower varieties, not the 12-foot-tall whoppers. Sunflowers come with so many faces today that there are plenty from which to choose.

Melons: I know of no reason why melons can't be grown in an unheated greenhouse, and I've heard of those who have done it. But I haven't. Just give it a try.

Tomatoes: These are a greenhouse staple, and we grow only beefsteak varieties because that's what customers demand early in the season. It's later on that they'll want paste tomatoes and cherry varieties (which we stopped growing because they take so long to harvest). Tomatoes need extremely rich soil for greenhouse production, as do peppers.

We grow hybrids for earliness (Early Girl is a favorite). For a main-season crop we only grow open-pollinated, indeterminate varieties, favoring the heirlooms for taste and seed-saving. Heirlooms keep gaining in popularity and customers ask for them.

Staking isn't mandatory in the field, but it is indoors, so we suspend twine from a purlin and train the plants to a single stem, pruning all "suckers" and wrapping them in a spiral around the twine as they grow. We space tomatoes at about 24-inch intervals, and start seedlings about five weeks before transplanting—about a month before last-frost date. You simply cannot hurry tomatoes; they need heat to grow.

Most fellow growers I know train their tomatoes on T-stakes with so-called "Florida weave" twine to support them. There's no right way. In the field, we grow tomatoes along a stock fence at the property's edge, but not everybody has such a convenient trellis.

Hoop house lettuces are most efficiently grown in wide rows with narrow paths between. A bed three feet wide or less allows you to step over it. Courtesy Eaton's Creek Organics

Green onions, or scallions, are one of spring's first crops.

Peppers: We've found that green bell peppers are still the favorite in terms of steady sales, but yellow and orange peppers are also in demand, and can be sold at a premium. (Check the difference in price at a supermarket next time you're there. It can be a dollar a pound or more.) Purple and white varieties might be worth a trial balloon, but they're a bit unusual and food shoppers tend to be fairly conservative.

Mexican food is becoming wildly popular, so we grow an increasing amount of both mild and hot jalapeños, cayennes, and chiles. In 2006 we tried, and liked, Tabasco peppers (the pepper of Tabasco sauce). These command a higher price than green bell peppers. We've tried growing habanera and other very hot peppers but stopped because (1) we sold few, (2) they burned our skin when picking, and (3) we don't do hot-hot.

Sow peppers in 2 3/8-inch cells four weeks before transplanting to the hoop house in mid-March (a month prior to frost-free date). The larger cells allow for a larger root system and earlier fruits. To push germination we keep peppers in the warmest part of the greenhouse, nearest the heater.

Space the plants 12 inches on center in 36-inch rows, and pick the moment the peppers reach good size. We price by the piece, not the pound, and have learned that large peppers earn a premium while costing us little or nothing extra to grow.

The Johnny's Selected Seed catalog relates that flower and fruit numbers can be increased in peppers by subjecting the seedlings to cool temperatures (53-55 degrees Fahrenheit) for four weeks in full sun, then return them to 70 degrees day and night before transplanting. Do this after the third true leaves appear, Johnny's recommends. I have no such controlled cool spot, and therefore have to pass on what sounds like a nice idea.

Others: Other crops that can be grown successfully in a cool house include radishes, green onions (scallions), celery (needs plenty of moisture), and summer squash. The latter is a crop we like to get to market as early as possible, because people are nearly as hungry for summer squash as they are for tomatoes, and spring supermarket squash can be about as unappetizing as winter tomatoes.

Mind you, if you're raising greenhouse crops in a cool house that has a heater, then it may be tempting to crank the heat up to get things moving along even faster. But that's self-defeating because the amount of heat required (unless you've got a wood stove in there with plenty of firewood) will be too expensive to make it worthwhile.

COOL-WEATHER PLANTS

Fall and winter are when the unheated greenhouse really comes into its own. My friend Eliot Coleman is renowned for pioneering the winter greenhouse garden, and can grow enough crops to make you wonder if they even *have* winter on his Maine farm. (Yes, they do. *Oooh*, yes they do.) For more information, I heartily endorse Eliot's two books on the matter, *The New Organic Grower* (Chelsea Green, 1989) and *The Four Season Harvest* (Chelsea Green, 1992).

Depending on your location, the winter harvest includes essentially salad greens and root vegetables, but it can also include late cabbage, broccoli, Brussels sprouts, Chinese cabbage, and others.

The first thing to know about growing in the winter greenhouse is that daylight is a premium. Lettuce sown in early September will be ready to harvest in 60 days (give or take), whereas a crop sown in early November will take twice that long. Early March sowings are back to 60 days. It's all a matter of light, and that varies considerably between the Deep South and the deep-freeze northern zones. The best thing to do is plant heavily

When we have any down time after harvest, we begin filling flats with inserts and potting soil.

Sweet Potato 75¢ lb.

Sweet potatoes are a good late season crop, especially in the South.

and regularly the first year, keeping every bed labeled and the harvest times noted, until you can plan in advance in subsequent years.

Start slow, though. It's easy to try to do too much, too fast in farming and get in over your head to where stress wears you down. If you put up two greenhouses, you don't have to plant every square inch of both the first year. Take time to learn.

Winter markets are also quite different from summer markets. Roadside stands won't work except in the warmest climates, and you don't want to be manning one in a Zone 4 winter even if you *do* get customers. A CSA can easily run through winter, and restaurants tend

to lust for good salad greens and other produce in the dark days of winter, so I'd try those first.

Plants you can grow in unheated winter greenhouses include arugula, beet greens, chard, chives, claytonia, endive, kale, lettuce (but not iceberg), mâche, minutina, parsley, radicchio, scallions, spinach, and turnip greens. The three hardiest are mâche, spinach, and scallions. Lettuce will stand up through light freezes, but not through winter in Zone 5 and above. Depending on location, any of these salad greens may be sown in succession from late summer into early fall. They will grow well in the beginning, then slow down as daylight wanes.

Above: Arugula is a hardy annual that makes a lively addition to salads, soups and sautéed vegetable dishes.

Left: In spring, when moving from the hoop houses to the field, some cleanup chores must be left until later.

This small hoop house was ideal for
selling tomato plants.

Alright, so you can't always get everything transplanted on time. Once a transplant is stressed, as these are, they aren't worth setting out.

With shade cloth and good ventilation, a hoop house makes an ideal spot to start midsummer transplants.

Other cold-hardy crops may also be sown in summer for late fall or winter harvest, in cells or direct-seeded. This includes (as mentioned above), cabbage, broccoli, Brussels sprouts, Chinese cabbage, and cauliflower. Brussels sprouts are especially hardy. None of these will grow much in cold weather, but they will keep a good while, running in place as it were.

Root crops are also good candidates for winter harvest. These can be planted in late summer, allowed to grow all season, and then left in the ground and harvested through winter. Carrots are my favorite, because soil temperatures in the mid-30s change some of their starch to sugar, making a carrot crop particularly sweet in colder weather. Outdoors, we mulch our carrot beds with leaves to keep the soil from freezing, and harvest (with the tops still green!) right through winter in Zone 6. In the greenhouse you can mulch the same way, or with any other insulation handy.

In raising these or any other greenhouse crops, the key to everything—your profits, your sanity, your marriage—is rich, friable soil. The rest is details. As stated earlier, the space demands on a greenhouse require intensive, sequential planting, and that in turn requires fertile soil. But you knew that.

In raising any greenhouse crops, the key to everything—your profits, your sanity, your marriage—is rich, friable soil.

Chapter 8

BUGS AND PESTS

Even with the 10-fold increase in insecticide use in the United States from 1945 to 1989, total crop losses from insect damage have nearly doubled from 7% to 13%.
—David Pimentel, Professor of Insect Ecology and Agricultural Sciences at Cornell University

Oriental eggplants are among the most beautiful of vegetables. Note the absence of many flea beetle holes in the foliage.

For generations we've treated farm crops as innocent victims, and insect pests as enemies. For generations we've declared war on these pests, going after them with all the poisons modern science could devise—spraying, dusting, burning, bombing, and attacking them until they submit (or not). For generations, we've treated bugs as we treat bacteria, fungus, or spiders and such.

And for generations, we've been going about it all wrong.

Bugs are part of our homes, part of our farms and gardens. They are part of life. They are wondrous creatures to behold and study. Sure, they may dine on our crops now and then. Plan on it. Expect it. Flea beetles will bother your eggplants; vine borers will hit your winter squash, and cabbageworms will go after your cole crops. But they won't destroy our farms. It doesn't happen that way. They come and go. But the richer our soil (and so the more healthy our plants), the more diverse our crops, the better we rotate fields, and the more balanced our farm's ecosystem, the less bugs will bother us.

Many years ago I made an interesting discovery about squash vine borers, and later learned I was not the first to know it. We'd been having trouble in the pumpkin patch with wilting leaves, so of course I looked for the telltale piles of brown frass near the base of the plants—and found them. I cut along the stem until I found the larvae, and dispatched them in each plant, one after another. Then I discovered that vines farther out from the base were not wilting at all. In fact, they were thriving and blossoming freely. I checked and found that the vining stems were rerooting themselves at intervals as they grew, thereby, in effect, outrunning the borer gnawing its way along.

The vigorous growth that results from planting on rich, fertile soil is the best defense, in this case because the vines grow quickly and produce a crop through secondary rooting. Fertilizing the main squash hill won't stop borers; but fertile soil 15 and 20 feet away will make them irrelevant. (Native Americans planted pumpkins and other squash among their corn, and while this can make large-scale harvesting somewhat more difficult, it does seem to repel, or simply confuse, the vine borer moths.) In either case, pesticides are the last thing you need.

CANARY IN THE COAL MINE

In fact, I find that insect "pests" are not so much a *problem* as they are a *symptom* of a problem—something is wrong with your soil or farm ecosystem. A good example occurred one year with Colorado potato beetles. We've never grown potatoes much as a cash crop because they aren't "sexy" enough to charge what they're worth in terms of our time spent planting, hilling up, harvesting, and washing. But we grow them for our own use on a small scale and so I can afford to hill them up—not with soil, but with rotting leaves or straw. As a result, potatoes grow in the richest, moistest, most crumbly and wormy soil you can imagine, and as a result of *that*, we have never been bothered by Colorado potato beetles. Oh, they come, they go, but they don't much hang around. Mind you, this used to annoy our neighbor greatly because his potato plants were infested with thousands of the beetles, causing no end of damage. When he finally asked, I finally explained what I did, and that he should try the same thing. He did, and it worked.

The problem was the soil, not the beetle.

Still, people come to me all the time, asking about "pest" problems, and the conversation often goes just like this:

"I have bugs all over my [beans, squash, roses, you name it] plants. What do I do?"

"What kind of bugs?" say I.

"I don't know," say they.

"Well what sort of damage are they doing?" say I.

" I'm not sure," say they.

A Black swallowtail on bee balm—the beauty of biodiversity. *Courtesy of Marie Jackson*

And there you have it. It's a bug, and therefore must be doing something dastardly.

My own experience, and those of countless others, is that stressed plants are susceptible to far more damage from herbivore insects than nonstressed, healthy plants. It works that way with humans, and it works with plants. Stress is a killer.

In plants, the stress may come from the lack of a simple micronutrient, a nutrient imbalance, excess acidity in the soil, lack of rain, or any number of other problems. But in every single case, without exception, the problem can be ameliorated or eliminated by incorporating organic matter into your soil from a broad range of sources.

THE ANTIOXIDANT CONNECTION

The relationship between organic plants and pests is remarkable when you consider the role of antioxidants. Researchers are discovering that these health-promoting compounds are far more prevalent in organic produce than in conventionally-farmed food, and several studies indicate that organic plants produce extra antioxidants as a defense against pests. Antioxidants apparently don't taste good to bugs. Conventional crops are provided pest protection by chemical insecticides, so they don't need to manufacture the same level of antioxidants as organic plants.

Several studies have found levels of specific vitamins and antioxidants in organic food to be two or three times the level found in matched samples of conventional foods, according to Charles Benbrook, Ph.D., chief scientist for The Organic Center for Education and

Promotion. Other studies provide evidence that core practices on organic farms, such as the use of compost, cover crops, and slow-release nitrogen, can increase antioxidant content in produce over that found on chemical farms, Benbrook says.

INTEGRATED PEST MANAGEMENT

Mind you, building up the health of your soil is a long-term task; it can't be done overnight. In the meantime, if any of your crops are being subjected to an inordinate amount of insect damage, or you fear they might be, the best approach to the problem is known as integrated pest management, or IPM.

This is a relatively recent concept that fits hand-in-glove with organic farming. What it means is that when a crop is threatened by an insect pest, pesticides (organic or otherwise) should only be considered a last resort. Instead, control bugs in ways that are least injurious to the soil and environment, and are most specific to the particular problem pest.

Some good examples are:

Plant pest-resistant varieties. Butternut squash, for instance, is far less apt to be bothered by vine borers than hubbard squash. Some, in fact, plant blue hubbard squash as a trap plant to lure the borer moths away from other cucurbits.

Use physical barriers such as floating row covers.

Plant a crop that will ripen before or after bug infestations are likely.

Use crop rotation. This is useful for lowering the density of both pests and weeds.

Diversify your crops as much as possible. If an infestation hits one crop, you'll have plenty more to sell.

Encourage beneficial wildlife such as birds, ladybugs, and others (See sidebar).

Use nonchemical pesticides when absolutely necessary, applying them in such a way as not to harm beneficials.

ORGANIC PESTICIDES

A complete list of pesticides approved for use on organic farms is available on the Organic Materials Review Institute website (*omri.org/FAC.html*). But in decades of growing for home and market, the only pesticides I've ever needed from this list have been *Bacillus thuringiensis* (or Bt, sold under various trade names), and insecticidal soap ("Safer" is one of the brand names). There are many others, including reasonably

Cabbage may be treated with Bt to prevent cabbageworm damage.

popular neem oils and pyrethrins (or *pyrethrum*, as the OMRI website spells it), but Bt and insecticidal soap are the only two I've found much cause to use.

While these are judged safe both for humans and the environment, it's still wise to use any pesticide in such a way as to not harm beneficials, worms, bees, or other life forms. Bt degrades in sunlight, and so should be applied only in the evening or on cloudy days (when honeybees are less active, coincidentally).

And in all cases, read the label carefully, using only as recommended. If a little is good, more is *not* better.

COMMON CROP PESTS AND TREATMENT

Just as I've only needed two pesticides in 30 years as a grower, the only serious insect pests I've come across in that time are the imported cabbageworm, the flea beetle, the Mexican bean beetle, vine borers, and the corn earworm. There are plenty more, mind you, such as the tomato hornworm and cutworm, and we could get into this and never get out, but for the sake of letting you get back to the soil, I'm going to stick with these and maybe a few more.

Imported cabbageworm: If you see white butterflies dancing through your cole crops, within a few days you'll be hit with an infestation of cabbageworms. These velvety green larvae don't much bother seedlings, but once a crop begins to head, damage can be immediate and severe. The most common controls are floating row covers as a physical barrier, and spraying or dusting with Bt. Both are highly effective.

In the late 1800s, growers often went after these butterflies with nets, and I've

Honeybees are one of the most essential beneficial insects on any farm, and can be done in by the same pesticides that kill other insects. *Courtesy of Delvin Farms*

been tempted to try that myself, but I think you must need some sort of net that I don't have—a fishing net won't cut it.

Corn earworm: These I've been able to beat by enriching the soil. It took some years, but in the end we had little problem. And we found, at our farm stand and in our farmers' market, that while some customers objected to a little worm damage at the ear tip, others didn't mind a bit. We'd sell 'Silver Queen' for 50 cents an ear and get it. It was that good.

The moth that lays its eggs on the corn silk usually comes at night, and therefore isn't seen laying its eggs on the fresh silk. Bt can be used as a control, but timing is difficult. Look for small greenish larvae with dark heads feeding on the ear tips, and spray immediately with Bt before they grow and burrow into the ear. Early or midseason crops are less apt to suffer than later ones, but not always.

Some suggest planting a sweet corn variety with tight husks as a sort of physical control against worm penetration, but I love 'Silver Queen', so we enrich the soil instead.

Flea beetles: These little beetles that jump like fleas when disturbed can decimate a young eggplant crop with their tiny leaf holes in no time. I've rarely had trouble with them elsewhere—just a bit with broccoli—and found that a few applications of organic insecticide usually stops them for the season on eggplant. Rich soil helps a

crop get through that vulnerable early stage quickly, and mature plants are bothered little or not at all.

Floating row covers applied early also help, and a trap crop of Chinese 'Southern Giant' mustard (*Brassica juncea* var. *crispifolia*) is said to make the beetles drop what else they're eating and run to that. Chinese daikon and 'Snow Bell' radishes are also trap crops for the flea beetles.

Mexican bean beetle: These can do serious damage to a bean crop, but we've found that most of the damage is caused after the crop is picked. The best defense is rich soil (which results in enough vegetation for both the beetle larvae and us) and planting an early crop, which is harvested the minute it's ready.

Others recommend the use of floating row covers applied soon after germination, and before the beetles (which look like, and are, a member of the lady beetle family) are laying their yellowish egg masses on the undersides of bean leaves. Neem oils are also said to be effective, though we've never used them. An undersown crop of clover or something similar has also been said to confuse the moths of this and other larval pests.

Vine borers: These creatures emerge from eggs laid near the base of a winter squash plant and begin to eat their way into the squash vines, causing leaves to wilt and the plant to, perhaps, die. The home

In one year, the average American consumes more than 100 pounds of food additives.

Salvia, in this case 'Blue Bedder', not only attracts pollinators, but is an excellent cutting flower.

"Of all insect species, more than 97 percent of those usually found in the home landscape are either beneficial or innocent bystanders."

off. Ever since we began hilling potatoes with straw and leaves, the problem has disappeared. If that doesn't work, or you don't have sufficient straw or leaves, then try a Bt variety known as *San Diego* (BTSD it's called).

Again, landscape contractors bring us their leaves. We don't grow straw, but found a farm that does, and every year it has maybe 20 tons left over that got wet and spoiled. So we're happy to help dispose of it. For manure, we pick it up free at the Nashville Mounted Patrol stables.

So look around, ask around, maybe put an ad in the paper. You'd be amazed how much free stuff is out there if you can haul it.

Beneficial insects:

As further reason not to wildly dispense pesticides on the farm, keep in mind that of all insect species, more than 97 percent of those usually found in the home landscape are either beneficial or innocent bystanders, according to the Maine Cooperative Extension Service.

Attracting beneficials

Many beneficial insects, in their adult stage, eat pollen and/or nectar. To attract them, pick annuals, biennials, and perennials that bloom for a long period. Small-flowering plants are particularly attractive to tiny parasitic wasps.

The list of beneficial-loving plants includes parsley, dill, fennel, members of the mint family (lemon balm, spearmint, thyme), daisies, coneflowers, yarrow, and goldenrod.

Other common nectar-producing plants include asters, bee balm, black-eyed Susans, borage, buckwheat, clover, cosmos, Joe-Pye weed, lavender, marigolds, onions (and onion family members such as chives and garlic), raspberries and other brambles, sage, marjoram, and tansy. There are plenty more. The more diversified your land, and neighboring land, the more likely you are to draw in beneficial insects.

gardener has the option of seeking out and killing individual larvae, and helping reroot the stems with a shovel full of damp soil. But with a field crop of pumpkins, this isn't an option—you need to keep your day job. We've found, honestly, that vine borers are one of those things you can generally afford to ignore. We've never lost much of a crop to them, and organic farming has nothing to do with losing a plant or two; it has to do with seldom losing much of a crop.

Cutworms: These I put in much the same category as vine borers. They do damage now and then, but not much, and I can't recall ever losing a direct-seeded plant to cutworms. They can take a heavy toll on transplants, but if I see a problem coming I scatter grass clippings on the area. This seems to confuse the poor critters. (Local landscape contractors bring us their clippings from untreated lawns, so we have plenty.)

Colorado potato beetle: Same thing again: A few of them come, chew, and take

The following is a list of the top 10 beneficial insects for agriculture:

1. Spiders: North America has some 3,000 named species, and most kill insects with venom. Most rely on a web, but others, such as jumping spiders and wolf spiders, go looking for trouble. A 2002 study found that spiders as a group are the most abundant pest predators on a wide range of plant material. So be nice.

2. Lacewings: The larvae of both green and brown lacewings are voracious predators of aphids, mites, thrips, and other small, soft pests.

3. Lady beetles: These "ladybugs" are equally predacious against aphids, but they also eat insect eggs, mealybugs, mites, and other soft-bodied insects. Flowering, pollen-producing plants attract them.

4. Ground beetles: These nocturnal predators (brown, black, or blue-black) have jaws to kill caterpillar pests, including armyworms, cutworms, grubs, snails, and slugs. Both adults and larvae are predators.

5. Praying mantis: These large, green, highly destructive pest predators may be up to 3 inches long and will feed on anything they can catch, including each other.

6. Predatory wasps: This category includes yellow jackets, paper wasps, and bald-face hornets, all of which are important predators of caterpillars and other soft-bodied insects.

7. Parasitic wasps: Insects in this group are generally tiny, black, and less than an eighth of an inch long. They attack a wide range of pests including tomato hornworms, cabbage loopers, imported cabbageworms, corn borers, leaf miners, and more by laying eggs inside the host bug. The egg hatches, and the larva eats its host from the inside out. Yucky, but effective.

8. Tachinid flies: This is a diverse group of more than 1,300 North American species, all of which lay eggs on the host insects. The larvae bite into the host to feed on it. Victims typically include caterpillars and beetles.

9. Predatory bugs: This group includes big-eyed bugs, damsel bugs, minute pirate bugs, and predaceous stink bugs. (The latter looks like its plant-feeding relative, but has a distinct spike on each side of its "shoulders.") This group dines on caterpillars, insect eggs, leafhoppers, beetle larvae, and many other types of insects.

10. Ambush bugs: These, like praying mantises, have enlarged, modified front legs, but are bright yellow—in part because they love to lie in wait in goldenrod flowers.

Enemies

"One of the most time-consuming things to have is an enemy," E. B. White once wrote.

It is also the dumbest, costliest, and among the hardest things in life to walk away from.

My worst enemy once was Lena, the Nubian milk goat. She was not only sneakier than I am (intelligence, in an enemy, is called sneakiness), but also was stronger, louder, more ornery, and—like any proper enemy—the clear cause of most trouble around the farm. Mind you, I was no small part of her troubles, but it's hard to have or keep an enemy when you look at both sides of things.

We bought her because Nubian goat milk is one of life's finer, lesser-known pleasures. Unfortunately Lena thought so, too, and had no desire to part with the stuff. Locked on her milking stanchion with a bowl of grain, she was fine. But one touch of a hand on her udder would bring an electric reaction. She'd kick, stomp, thrash, and buck her hindquarters so high that now and then her left rear leg would swing over a 3-foot-high plank and leave the poor girl hanging there, stuck fore and aft, half upside right, half down, howling in protest with the notorious, other-worldly "BLAAAAAAAA!" that only Nubian goats are capable of. We know for a fact she could be heard down the road at Tom Merna's place, and hearing the call inside a closed milk room is a particularly special experience.

By and by she'd calm down and allow me to milk until the pail was half full, at which point, without warning, a back foot would shoot and hit the pail's lip like a tidily wink, redecorating me and part of two walls with milk. In response I tied her rear feet to eye hooks screwed into the milk stand. She considered the situation and resolved it by lying down on the milk pail. I counterattacked by slinging a grain sack under her belly and hoisting her midsection up with rope and a block and tackle. That left her teats pointed due east on the horizontal, however, which raised accurate milking to an art form, in which one must calculate volume, trajectory, and distance if you're to get enough milk for morning coffee.

Lena is retired now, a milker emeritus, but when her last two kids grew to half her size they'd stare at me now and then when I came to fill the water trough. They knew what was ahead and were waiting, preparing, taking my measure. In the meantime their mother kept up the war by wearing a new hole in the pasture fence every week or so. I could have bought a high-tensile electric fence, but the people who make them hadn't met Lena. Besides, that wouldn't end the conflict; it would merely raise the stakes and lower my savings.

Like most wars, this one was senseless. But having declared it and come this far, neither one of us was inclined to walk away. Once in a while I'd scratch Lena behind the ears, or rub the bridge of her nose, and she liked that. Maybe I should have done it more often. Maybe she needed a larger pasture.

My latest enemy was the local squirrel population, which had grown greatly. That put acorn storage space at a premium, and it overflowed onto our front lawn. (A gray squirrel buries each of its nuts and other winter stores well away from other squirrels' pantries, lest conflict erupt at gathering time.) This made a mess of the lawn, so I thought to solve the problem with a .22, randomly picking off the first of what I thought would be many squirrels. I shot it off a tree branch with a long, lucky shot. The acrobat of summer twitched slightly, and leaned, then plummeted dead to the forest floor.

I didn't like the feeling, and decided that perhaps the real enemy here was me.

Chapter 9

WEED LIMITS

ere's the bad news: One lamb-squarter weed produces about 70,000 seeds. Pigweed produces 100,000 or so. Weed seeds can lie dormant in your soil for years, some for decades. The vertical roots of field bindweed can grow 20 feet deep. A study of cropland in western Nebraska found 200 million seeds per acre, or 140 seeds in every pound of soil (about a big handful).

That's what we're up against.

Weeds have been the bane of growers since hunter-gatherers first settled down to plant cereal grains in Jericho 10,000 years ago. Cultivation was the best control for most of those 10,000 years—until we fell into the easy arms of herbicides. The British discovered the chemical herbicide, 2,4-d, in World War II, and ever since then we've been offered chemicals with such testosterone names as Eliminator, Roundup, and Zap-It. These aren't packaged as weed *controls*, mind you, they're packaged as killers. Weed *killers*. Another enemy, just like bugs. Organic growers don't use these herbicides, so what can we do? Are we back to where we were 50 years ago?

When faced with a dilemma like this, I often stand back and ask myself, "What's the goal here? What's our real goal?"

In this case, is our goal to have weed-free soil? Well, sort of, but not really. Our *real* goal is to have abundant crops. And if that's the case, we can begin easing our way into the good news here by stating flatly that organically raised crops withstand weed competition far better than conventional crops. You can have weeds *and* abundant, high-quality crops at the same time, it turns out.

That's been my experience for years, but it's also the conclusion of a 2005 study done by the Rodale Institute on the effect of weeds on corn and soybean crops. This was part of the institute's Farming Systems Trial, which is a long-term comparison of organic and conventional farming practices in Pennsylvania.

"Weed competition appears to be much less in the organic plots than in the conventional plots," conclude the authors of a report on the study.

"At the Rodale Institute we occasionally have a field or two where weeds develop prodigiously. This is usually a result of poor weather conditions that prevent normal (cultivation)," the report goes on. "According to the Penn State agronomy guide, these corn crops should suffer massive yield losses (of roughly 40 percent). Interestingly, however, in most of these cases our corn crops still yield competitively."

The reason for this is the fertile soil found in well-managed organic farms, the report concluded. Conventional farms treat the soil as a planting medium; something to hold plants in place until the crop ripens. Fertility comes from applications of chemical fertilizer. To organic farmers, on the other hand, the soil is a resource, a living thing to be fed and nurtured and nourished into good health in its own right. The result is healthy, abundant crops. One follows the other as night follows day.

According to the institute, research has shown that weeds take up chemical fertilizers quicker and in larger quantities than do field crops. As a result, these weeds provide powerful competition early in the growing cycle, putting severe stress on both the farmer and the crop plants at a crucial time. On organic farms where the soil is rich in organic matter, nutrients break down and are made available to plants on a slower, more long-term basis, giving weeds no head start, and no opportunity to deplete nutrients at a critical time for field crops. In addition, since soil rich in organic matter holds moisture better during drought and has a deeper reserve of plant nutrients, even if weeds thrive, the soil is able to nurture both them and the field crop. There's room for many at this table, it would seem.

That isn't to say weeds aren't a problem. They are, and there is much we can do to limit their success on an organic farm. It is only to say that an organic crop can apparently tolerate a higher weed threshold than a conventional crop without losing significant quality or abundance. That threshold at the Farming Systems Trial was 2 tons of weeds (dry weight) per acre. After that, corn yields diminished. But hey, 2 tons *dry weight* per acre? That's a lot of weeds.

WHAT THEY DO

The best way to limit weeds is first to understand thoroughly how they grow; to know their patterns and weaknesses. Much has been learned about this in recent years, in part due to the burgeoning popularity of organics and the growing awareness of chemical hazards.

Besides just looking shabby, weeds reduce crop yields and quality in three ways:

One, they compete with a crop seedling for sunlight. Some annual weeds grow extremely fast and can shade an emerging crop, thereby reducing photosynthesis, and ultimately growth. In effect, you can't find the corn in amongst the Johnson grass.

Two, weeds compete with crops for water—and they're often much better at finding water than crops. What, after all, will you find growing up out of a crack in the concrete—plantain or broccoli? Weeds love dry, open soil; that's their niche. They are like Brer Rabbit in the briar patch. Organic growers often transplant into black plastic (or in summer, white, to reflect light) with drip irrigation beneath the plastic as a water source. But this water doesn't remain under the plastic for long; it readily and constantly wicks out to the dry soil on either side. "A cup of rain is worth a gallon from the well," the old saying went, and it's true. So use plastic to keep weeds away from your row crops, but don't expect this to make weeds suffer much between the rows.

Three, weeds compete with crops for soil nutrients. As mentioned, rich organic soil may have enough nutrients to feed weed and crop in the short term without the crop suffering much (in the case of corn, a heavy feeder), but why look at the short term? Every weed that goes to seed is compounding the problem next year and for years to come. Weeds are long-term.

Finally, weeds can be an eyesore. Whenever I visit another farm during growing season, the first words out of the farmer's mouth are usually, "We've got nothing but weeds." Virtually every farmer or gardener I've ever known has a deep sense of aesthetics, after all, and they want the place to look good. That can mean lush rows or beds of crops, and spanking clean soil between them.

Unfortunately, many people equate a lack of weeds with skill in growing. I had a woman working for me once who loved to weed. She loved bare earth. She loved it so much in fact that I swear she would have cultivated every weed and crop plant alike

Research has shown that weeds take up chemical fertilizers quicker and in larger quantities than do field crops.

if it meant being able to look out on a sea of brown earth. Such an attitude leads conventional farmers to overuse herbicides, and I know can cause organic growers to get discouraged.

I'd be tempted to suggest that one find a good balance between crop loss and excess land management, but the plain fact is that no farmer I know has enough time to over-cultivate, even if they wanted to. The balance they strike is the one their experience and instincts dictate.

HOW WEEDS WORK

Weeds are nothing more than nature's attempt to bring stability to what she considers a highly unstable and volatile environment. That is to say, a tilled field. Nature doesn't want this. Nature wants a mature grassland or woodland with a high degree of biodiversity. What *we* want, on the other hand, is a high degree of corn or broccoli.

And so, into our tilled field the weed seeds come, carried by birds, by us, by our tiller, by rain runoff. In the first year after a piece of fallow land is put to crops, not many weeds are evident because the land was, in some sense, stable. But don't be lulled. Down the road is when you run into trouble.

Annual weeds are well adapted to grow in areas subject to periodic disturbances such as plowing or tilling. To survive, they must grow to seed before the next disturbance, so annual grass weeds grow quickly and put a good part of their life cycle into making seeds for another generation. Annual plants must also produce more seeds than perennials because that's their defense against being uprooted. Their seed is apt to be long-lived, and annual seeds have dispersal systems that are all over the map, from burrs on your dog to little flying parachutes arriving on a wind from your neighbor.

Perennial weeds prosper in more stable environments such as hay fields and no-till cropping systems. Perennial weeds have long-lived seed just like annuals (although producing less of it usually), and also may be able to regenerate without seeds, such as by the underground rooting systems that make such a nuisance of quack grass (or witch grass), Johnson grass, and Bermuda grass. These stolons, tubers, and rhizomes also store food for the weed so it can regrow rapidly when chopped up or disturbed.

All three (and other perennial weeds) also reproduce from seed, usually by the first week in July in Zone 5, so each must be uprooted before that. Johnson grass is a common hay weed, so in any area where it's common, never, never, mulch with hay or use it in anything but a hot compost pile.

THE KEYS TO CONTROL

Rather than simply dealing with weeds by cultivating, mulching, or applying herbicides, it helps greatly to step back a bit and look at the problem as not so much "How do I get rid of these weeds?" but rather, what can you as an organic farmer do by design to prevent the weeds from coming, or even more importantly, to make them less relevant.

1. Drain the Weed Seed Bank

The weed seed bank in your soil amounts to the reserves of viable weed seed that will be able to germinate under proper circumstances. This includes seeds recently shed, as well as those that have been in the soil for years, sometimes decades. They can lie in wait just under the soil surface, or a foot deep, and everywhere in between. Less than 10 percent of these weeds will germinate in a given year, so the seed bank can be considerable.

The chance of these seeds germinating depends on many factors, and an important one is how deeply the soil is disturbed each year. Seeds buried close to the soil surface are subject to predation by birds, insects, rodents, weather, wind, and more. Those buried deeply, however, are buffered from extremes of temperature and moisture. Weed seeds dug up from a depth of 10 inches germinated at rates of 34 to 38 percent, according to one recent study, whereas those buried only an inch deep showed only a germination rate of less than 5 percent. This seems to suggest that, in terms of weed seed germination at least, regular shallow tilling is more effective than periodically disturbing the soil with, for example, a moldboard plow.

2. Cultivate

Cultivate solely to keep weeds from going to seed, and do it when the weeds are just past germination. That is, at their "white thread" stage. And cultivate before, not after, a rain. Since the weeds are small, even tiny at this stage, it's tempting to wait until they become an eyesore. But wait until then and you'll be weeding and chopping, not cultivating, and that's a much tougher chore, clogging up both machinery and your day.

With all peas and beans, I broadcast thickly in a wide row to choke out weeds and make the most efficient use of small acreage. Most all of our soil is cropped or mulched; little is left open for weeds. We also plant all root vegetables and salad greens in wide beds.

Where weeds such as quack grass or Johnson grass have taken hold, spring-tooth harrows (or something similar, such as a cultivator or spader) can be used to pull the rhizomes to the soil surface where they dry out and die. Discing and tilling, on the other hand, tend to cut up the roots into pieces which become new weeds. You want to pull them up,

Plastic mulch over drip tapes makes for excellent weed control around the plants, but not between rows. White mulch is used when no solar heat gain is required, as one gets with black plastic.

not chop them up. Subsequent cultivation should be shallow; no more than an inch.

On smaller farms, I recommend cultivating with a stirrup-bladed wheel hoe or shallow power tiller. Wheel hoes are ideal for narrow rows, and the blades range in widths from 6 to 14 inches. A one-wheeled model with two stirrup "wings" can cultivate a width of 32 inches. Fitted with two wheels, the hoe can straddle one crop row and cultivate two rows at once. But, once again, the weeds must be small. These cut weed seedlings down without turning up more seeds.

Field trials and my own experience have shown that if weeds are prevented from setting seed in a field for four years, then that field's seed bank is pretty well depleted. More seeds will come, mind you, and more shallow cultivating will be required, but the monster's back will have been broken.

3. Mulch

This is far more feasible on a small scale than on large operations because I know of no machinery that can do it. We use a pickup truck and yard cart to cover up every square inch of dirt we can find, and what we don't mulch we plant with a cover crop. The advantage of mulch is its multiple purposes in:

(1) adding a great amount of organic matter to the soil as sheet compost,

(2) preventing soil moisture evaporation,

(3) smothering weeds with a blanket of darkness,

(4) preventing soil-borne diseases or dirt from splashing up onto a row crop, and

(5) providing a nice soft, clean surface to walk on or kneel upon while fertilizing or harvesting.

We discovered this year for the first time that a combination of tillage and immediate mulch with oat straw prevents virtually all weeds from appearing in our clay loam soil. And the layer of oat straw needn't be thick; no more than an inch or so. We discovered this after harvesting beets and pulling up the straw alongside for a second planting. It took a moment to see, but not a single weed, including my nemesis Bermuda grass, had come up through the straw in more than two months. On bare ground barely inches away, however, weeds had taken over completely. We'll test this next year to see what happens.

4. Prevent Recurrence

Once the seed bank is under control, don't allow it to reseed from weeds on the edges of your property or between fields. Keep these areas harrowed or mowed whenever possible. If tillage equipment has been used in fields with a bad weed problem, wash it well before using it on other fields.

Perhaps even more important, keep weed seeds from sneaking in through mulch or manure. We once mulched and composted successfully and heavily with spoiled hay, so long as we knew the field it came from and what grew there. Having no source of hay, we use wheat straw from a neighboring farmer who sells baled straw. Last year he had about 20 tons of spoiled bales left over for our use. I'd hug the fellow if he weren't so ornery.

Manure from any hoofed livestock is apt to contain weed seeds, so this must be avoided entirely or hot-composted to kill any weed seeds (see Chapter 6 on composting). It won't kill all, but it can kill most weed seeds, and farming is not about perfection. Chickens and turkeys don't pass weed seeds through their systems, but we compost this anyway to "cool" off the high nitrogen content.

5. Diversify Rotations

To keep a certain weed from establishing itself or spreading in a field, try alternating your crops to have various germination times and tilling requirements. Nearly all studies done on the relationship between weed density and crop rotation indicate that rotation is a valuable practice for this reason alone. To break the life cycle of weeds, try alternating between cold- and warm-weather crops, legumes, cover crops, and crops that can be controlled mechanically. Vegetables such as carrots, lettuce, and onions, which are slow to grow and don't produce enough foliage to shade out weeds, can be sown in an area previously sown with cucumbers, sweet potatoes, or some other crop that provides natural weed control. (In sowing carrots, mix in radish seeds so you can identify the row and cultivate close to it after the radishes germinate in three or four days. Close cultivation will bury the radish seedlings, allowing the carrots free rein.)

Market growers sowing a wide variety of crops will have an easy time of crop rotation, and the possibilities are endless. Even a simple rotation of legume/corn/vine vegetable will not only suppress weeds but improve insect and disease control. Since different vegetables take different nutrients from the ground, and take it in different ways with different root structures, a good rotation system will take full advantage of a soil's wealth. There is no other way to go.

6. Use Cover Crops

These not only choke out weeds with a dense canopy, but can add nitrogen and organic matter to your soil while improving tilth. Select species for rapid growth. Sow at high rates and irrigate if necessary to ensure quick, thick stands.

Buckwheat is an excellent hot-weather crop, not only for shading out weeds but

because it is allelopathic to witch grass, pigweed, and other weeds. The responsible compound is found in the stems, so its toxicity to weeds is presumably most active after the buckwheat is harvested. Winter rye also contains allelopathic compounds that prevent weed seed germination, but it's likely that rye's rapid, smothering growth is also responsible. Buckwheat and sorghum/Sudan grass smother weeds by outdistancing their growth.

The use of cover crops also reduces the need for frequent cultivation, which not only brings up new weed seeds but can harm soil structure and speed up the decomposition of organic matter in the soil.

The benefit of cover crops to farm soil is so great that they are often referred to as "green manures." They prevent erosion, move nutrients from lower to upper layers of the soil, add organic matter to the soil (thereby enhancing tilth and drought resistance), and keep nutrients from leaching away.

Some cover crops, such as winter rye and hairy vetch, may be rolled down to form a nearly impenetrable mat that will suppress weeds while preventing erosion or soil moisture loss. This must be done only after the crop reaches full flowering, lest the crop simply bounce back. The Rodale Institute recently contrived an ingenious system for rolling and planting a field in one pass with a front-mounted roller on a three-point hitch, and a rear-mounted drill. (Search for "crop roller" at www.newfarm.org.)

7. Undersow and Intercrop

Undersowing refers to the practice of growing a green manure, such as red clover or soybeans, in with a crop of corn. In effect, it combines a cover crop with a market crop on the same field. This provides the multiple benefits of weed suppression, companion planting, soil improvement, and moisture retention.

Intercropping is much the same thing, except that it involves the simultaneous planting of two or more market crops, rather than a cover crop. An intercrop of corn, beans, and squash has been around for generations because the corn gains sunlight and height, giving the pole beans a place to climb (while fixing nitrogen in the soil), and the squash captures light filtering down while shading the soil against weeds. A study in Tabasco, Mexico, with the inter-planting system resulted in almost a 50 percent corn yield increase over corn being

Whenever possible, get produce up off the ground at a farmer's market or farmstand. If a bushel basket is half empty, replace it with a peck basket.

grown alone. Bean and squash yields were severely reduced (compared to growing alone), but if corn is your goal, then the system works wonderfully.

Besides, we're talking about weed suppression here, not bean yields.

8. Flaming

Flaming amounts to killing weeds with the flame of a propane burner that is designed specifically for such a purpose. The tank and burner may be carried back-pack-style, pushed, or tractor-mounted, and are generally designed for preemergence use. Flaming can be ideal for leeks, carrots, and other slow-germinating crops.

Preplant flaming is also referred to as the "stale seedbed" technique. After the bed is prepared for planting, weed seeds (most in the upper 2 inches of soil) are allowed to germinate and emerge. Assuming adequate moisture and soil warmth, this takes 10 to 14 days. The weed seedlings are then seared with the flamer at their most susceptible growth stage, and the crop is seeded as soon afterward as possible.

The searing doesn't mean burning; it means heating weed seedlings for just the second or so it takes for plant cells to heat up, expand, and rupture. If you press a leaf lightly between your fingers and see a dark fingerprint, the job is done. Hot, dry weather seems to provide better results.

Another technique is to prepare the seedbed, drill the crop, and then flame just before the crop emerges. I've heard of European growers who place a pane of glass over the crop row to push seed emergence by a day or two. When those early seeds appear, the bed is flamed. This works well with slow-germinating seeds, such as carrots, which would find more weed competition if drilled after flaming.

Keep in mind that the growing tip of some grasses (including corn) are still below ground when the plant is an inch or two tall, so flaming won't work on these. It's nice on broadleaf weeds, however.

Propane prices make flaming something of a costly weed control measure, but it's worth considering as one tool in a big medicine cupboard of cures for limiting weed problems.

9. Please Do Not Feed the Weeds

Whereas irrigation water will leach off into dry, adjoining soils, fertilizer won't. Well, not much anyway except after a rain.

This means that when fertilizing a row crop, band-fertilize no more than 3 or 4 inches from the plants. Don't broadcast.

10. Tighten Up

One of the easiest techniques for weed control on fertile soil is to increase your crop seed density. This allows the crop to intercept enough light so that less than 50 percent of it will reach the soil surface, eliminating most weeds right there. We've done this for years with root vegetables, legumes, leaf greens, peppers, broccoli, cabbage, and corn. Come to think of it, the method has proven itself so successful that I wish I hadn't saved it for last.

This handy device will prepare a bed, lay drip irrigation tape, lay down plastic mulch, and cover its edges in one operation.

Chapter 10

TRACTORS AND TILLERS

The decision on tractors can be a tough one. These two vintage machines may never work again because the farmer found he spent far too much time repairing them, and bought a new one. If you're mechanical-minded, and don't need a machine for that much work, one such as these may be fine.

My brother Michael, a doctor and farm owner in rural Michigan, will never own a tractor because of what he calls "The Joys of Griping."

"I'd rather complain about how much work it will be to move this or mow that, and how much pain I'd be in afterward," he explains. "A tractor would ruin everything."

My friend Paul Ceccarelli takes a different view. First he bought a Toro Wheel Horse riding mower and snowplow for his three acres and garden. Then he traded up to a John Deere with hydraulics and a self-angling plow. Then he bought a 28-acre farm and traded up to a John Deere 855. Two years later he traded up again, this time to a John Deere 970 diesel tractor with power steering and four-wheel drive.

"Now I'm looking to buy 600 acres so I can get one of those *really* big tractors," he says with a straight face.

One helpful tractor attachment for a larger farm is an automatic transplanter. Of course, few things are *completely* automatic. *Courtesy of Delvin Farms*

Paul enjoys his machine. "I couldn't be without it. I can move stuff. I can pick up manure from the farm next door. I've got a harrow for the gardens and a brush hog for the seven acres out back. I can grade, plow, excavate, and dig holes. I can pull a hay rake and tedder for haying, and borrow a baler."

Next to a pickup, a tractor or large tiller is almost indispensable for farming a few acres of land or more, and it's merely practical. Some, like Paul, buy it for what it can do, not for what they need. Others wisely base a decision on what work needs to be done, how often it needs to be done, and how much of it there is.

If the snowplow cowboy next door will plow your drive for $200 a winter, you don't need a tractor. The same holds true for mowing. But if you're cultivating some acreage and your neighbor doesn't have equipment you can rent, borrow, or barter for, then you're in the market for a tractor. (Or a tiller; we'll look at those in a moment.)

To get what you need for the lowest possible cost, decide first what attachments you'll require before you look at the machine itself. (See sidebar on attachments.) After all, a tractor is nothing more than wheels, controls, and an engine. The attachments are what allow you to get work done.

If you need many attachments, which can get expensive, remember that it's common practice for farmers and landowners to share them (although in the case of balers and snowplows, everybody might need them on the same day). Still, if your neighbor has an implement you'll need occasionally, consider buying a tractor compatible with what's around. You will also need a machine powerful and versatile enough to handle these expensive attachments.

TRACTOR TYPES

When choosing a tractor, the first consideration is size—meaning horsepower, weight, wheelbase, and width. These four

A tractor or large tiller is almost indispensable for farming a few acres of land or more, and it's merely practical.

The indomitable disk harrow is invaluable for breaking up and smoothing soil.

Attachments

The following are a few common farm attachments to consider for your tractor: cultivator, tiller, disk harrow, tine harrow, manure spreader, bucket, cultipacker, fertilizer hopper, drills, brush hog, roller, chisel plow. That's only a beginning, because once you get in here it's hard to get out. On our place, we pretty much get by with a tiller, cultivator, bucket, seeder, and cultipacker. Oh, and a brush hog.

characteristics pretty much determine the scale of work the machine can do. A short, lightweight tractor can't lift or pull much of a load, for example, no matter what its horsepower is; and a narrow tractor with a short wheelbase is notoriously unstable when carrying a heavy or awkward load. Again, decide what you need done and how often it needs doing, then scale the tractor accordingly.

Compact tractors, or so-called garden tractors, are a step above riding mowers and generally cost less than $15,000. With the right attachments, these machines can handle many light-duty chores such as towing a trailer, mowing, wood-splitting, snowplowing, and tilling. Most have air-cooled gasoline engines, which cost less than diesel tractors initially.

Compact tractors aren't designed to last as long as their larger brethren, though, and it's not unusual to find an engine too powerful for the tractor's steering, suspension, and drivetrain systems. Horsepower

sells tractors but it can also kill them, because owners are fooled by the high horsepower and overstrain their machines. It's always better to oversize a tractor than to overestimate what it can do. Low prices make these small machines extremely popular for working a few acres, however, and many will last for years with proper care.

Utility tractors are the workhorses of the small farm, and are designed to last. They are heavier, larger, more powerful, and more versatile than compact tractors. They typically have far stronger frames and axles, less sheet metal and plastic, and can be repaired more readily. They cost more than $20,000 new, depending on options and attachments. Machines in this class are often traded in for larger sizes after a few years, so you're apt to find many used ones in excellent condition. Paul Ceccarelli's last tractor had two years and 86 hours on it (tractor "mileage" is measured in hours of use). He bought it for $16,000 and expects he could sell it today for close to that.

We cultivate less than five acres and have never needed more than a large rear-tine tiller, a small tiller (because Janet bought it for me), and a rugged but small John Deere 855 tractor.

NEW OR USED

When shopping for a new tractor, you'll find many manufacturers, some familiar, others not. In general, the manufacturers whose names you recognize are likely to offer good reliability and service. If you come across an oddball manufacturer, get

the names of a few local owners from the dealership and ask their opinions. Keep in mind that many people will swear by their machines because to do otherwise means they made a mistake, so ask for details.

To get the most tractor for the least money, consider buying a used machine. Nearly everyone does because good tractors are rugged and can last for 50 years or more. Check newspaper classifieds, the Internet, county extension agents, ag bulletins, and word of mouth to learn what's available in the area. Parts, manuals,

Mowing green manure before turning it under in preparation for another planting.

Small-scale farm seeders, such as this Earthway model, can be ganged together with 2x4s to make a three-, four-, or even five-row seeder for wide beds.

Be smart and never traverse steep land, even steep bumps in the land, horizontally—even if you've done it before, because a roll bar might not be enough to save you.

tools, and attachments are available for most old tractors with good reputations, but you'll have to be something of a "shade-tree mechanic" and pay close attention to maintenance. Used machines also come in a wide array of sizes, designs, horsepower, and versatility; don't expect them all to be big farm machines.

One drawback to older tractors is that they typically lack a roll bar, or rollover protection system (ROPS) as they're called in this age of naming everything a *system*. This feature is critical, in my book, because the combination of big wheels, high torque, and a high center of gravity makes tractors inherently unstable—and not just on hilly terrain. If a tractor's rear wheels are stuck in mud or frozen ground and you try to go forward too quickly, the machine can flip over backward and crush you. Fortunately, a roll bar (I still call them roll bars) can be welded on to older machines to prevent such deadly mishaps.

A word on this: Be smart and never

traverse steep land, even steep bumps in the land, horizontally—even if you've done it before, because a roll bar might not be enough to save you. My Uncle Tony rolled his tractor on his farm in Vermont on land he'd hayed a hundred times before, and emerged from the accident unscathed. Then he died of a heart attack the next day.

OPTIONS AND EXTRAS

As with attachments and implements, other options for a tractor depend largely on the work you need done and the conditions under which you'll be doing it. Four-wheel drive can be useful if you plow much snow or work regularly in muddy areas. It also tends to make a tractor more stable, since the rear wheels can be smaller, and thus the center of gravity is lower. Rear wheels on tractors are otherwise quite large to prevent them from getting stuck.

Power steering can be an arm-saver if you do a lot of turning on difficult terrain, or if you regularly carry a snowplow or loaded front-end bucket. A loaded bucket weighs

down the front wheels and makes steering difficult. The usual cure is to weigh the rear end down with a 50-gallon drum of concrete to increase traction on the rear wheels and counterbalance the front end to make it "lighter." Manual steering will usually suffice.

A three-point hitch is the triangle-shaped rear hitch that's pretty much a given on utility or larger tractors, and on some smaller tractors as well. The hitch has two hydraulically controlled lower arms, which lift, lower, and tilt a towed implement, and a free-swinging upper arm (or top link). It takes only minutes to attach an implement to a three-point hitch—unless your fingers

are half-frozen, in which case the operation takes longer and involves some cursing.

Three-point hitches vary in size and lifting power, from the category 0 hitches found on garden and compact tractors to the category III hitches found on the largest machines. Category 0 hitches can lift about 500 pounds, give or take; category I hitches found on utility tractors can lift two or three times that weight, and have to because their attachments are larger. Category I implements are easy to find, new or used.

A power takeoff, or PTO, is a removable extension of the tractor's drivetrain, and transfers power from the engine to remote

Having used this big old gangly thing, I learned to love it. But, it's old.

Used tractors can be a bargain in the hands of a good shade-tree mechanic.

No matter what precautions you take, and how many years you've used your tractor and implements, always approach things as if today's the day something could happen.

implements, such as hay balers or tillers, through a spinning drive shaft. You'll almost certainly need a PTO for farm work, if for nothing but tilling.

SAFETY MATTERS

The last words here should be foremost in your mind: Tractors kill. The rules for avoiding that are simple:

Have a ROPS and always wear a seat belt. If your tractor rolls over, you can't fall out and be crushed. If you own an old tractor with no roll bar, weld one on. If you have no ROPS, for some insane reason, then *don't* wear a seat belt; should your tractor roll, there's a slight chance you can jump free.

Have your PTO fitted with a nonrotating collar or shield. A drive shaft turning at 540 rpms is dangerous and you can figure out why.

Never allow passengers on a single-seat tractor.

No matter what precautions you take, and how many years you've used your tractor and implements, always approach things as if today's the day something could happen. Because it is.

TILLERS

If you're looking for a tiller for cultivating anything more than a large garden, don't look at anything but a large, rear-tined tiller with a 7- to 14-horsepower engine. Till widths on these usually range from 20 to 30 inches; they can weigh up to 500 pounds.

This might seem unwieldy if you're slight of build, but the fact is that a big rear-tine tiller is easier to operate and control than a medium-sized machine with front tines. These can bounce you to death. The big rear-

Diesel or Gas

The decision to buy a gasoline- or diesel-powered tractor is an easy one—sort of. Let's look at the weak and strong points of each.

Diesels are more fuel-efficient. Part of the reason is their high compression ratio, which means they can get more power from fuel. Diesel-fuel efficiency also stems from the fact that diesel fuel is heavier (denser) than gas, so you get more energy per gallon.

Diesels also have more lugging power—they can sustain momentary overloads without stalling, as might occur when plowing into a piece of low, muddy clay. A gas tractor in the same spot would have to be stopped and shifted to a lower gear before resuming, then stopped and reshifted again for the rest of the field.

Finally, diesels are more reliable and easier to maintain. Their engines, for one thing, are heavier and more rugged to withstand their higher combustion ratios.

On the downside, diesels cost a good deal more to buy, new or used, and cost more to repair. The fuel must be more carefully stored to be kept clean and water-free lest it cause the fuel injection pump to fail. The diesel is also harder to start in cold weather (but then you're not apt to be tilling frozen ground that day anyway).

Gasoline engines may be the best bet for a small farm because they cost less, and the amount of maintenance required for one or two tractors is not enormous. It's not ruinous in terms of time, and the replacement of parts is readily learned. The parts for a gas engine are also cheaper than those on a diesel. Winter starting is also easier.

When shopping for a used model, remember that many tractors are unloaded toward the end of a wear cycle when expensive repairs are coming due. Unless you know the tractor and its owner, it's wise to hire a mechanic to check out the machine and give you an estimate on upcoming repairs.

Whatever engine or tractor you buy, purchase the operator's manual as well (these are available for antiques), and stick to the recommended maintenance religiously.

The United States devotes about 4 percent of its energy use to packaging food. That's almost the amount we use to grow it.

tine machine with tractor tires has a front-mounted engine that balances the rear tines roughly over the axel. As a result, the operator needs to exert only minimal effort to guide the machine. Independent wheel brakes on some machines make them even easier to maneuver.

Some of these machines also come with attachments you might want to consider. These are changing constantly, so I won't list them. The only option I would definitely consider is an electronic starter. I hate starter ropes. Always have.

Chapter 11

TRANSITION TO ORGANIC

Fields used for haying can often be transitioned to organic immediately, whereas fields treated with chemicals require a waiting period to be certified organic.

The grower changing over from conventional to organic farming will find two very unfamiliar companions on his journey.

One is the need to completely change one's mindset about farm inputs. On a conventional farm, any problem has a quick-fix solution. Worn-out soils, pests, weeds, diseases—all could be dealt with by purchasing one or another off-farm input. On the organic farm, however, you don't react to problems, you invest in the future by building up your soil and heading off problems before they occur. This means having a personal relationship with your soil.

The second companion is the fact that the National Organic Program (NOP) has strict rules on what can and cannot be done or used on your farm if you are to be certified organic. This means losing a certain sense of control, but it also protects consumers, yourself, and of course the title, organic. In the end, in fact, the rules will turn out to be a help in converting to a sustainable system.

Conventional farmers making the change usually expect a drop in profits for several years as the soil gives up its addiction to chemical fixes, but this does not have to be the case, and often it's not. The first step lies in doing some research, and the first step there lies in understanding the USDA regulations.

The USDA defines organic agriculture as "a production system that is managed to

respond to site-specific conditions by integrating cultural, biological, and mechanical practices that foster recycling of resources, promote ecological balance, and conserve biodiversity."

This means, among other things:

1. The elimination of synthetic pesticides, fertilizers, and other inputs, along with hormones and antibiotics for livestock and poultry.
2. The reduction of off-farm inputs of any type.
3. The use of cover crops, manures, compost, crop rotation, and other practices to promote soil health.
4. The use of rotational grazing and mixed-forage pastures and alternative health care to promote animal well-being.
5. The promotion of soil health, crop rotations, and mechanical means to control weeds, disease, and insect pests.
6. An emphasis on biodiversity, soil and water conservation, and ecological balance.

Organic farming does not mean substituting Bt for a synthetic pesticide, or using fish emulsion as a fertilizer instead of chemicals. It's all about changing your entire way of thinking about plants, food, and the environment. Organic farming is about going *with* the natural world, on its own terms, and not trying to control it. It's about patience, observation, a new respect for the health of your customers, and a willingness to learn all over again.

Above all, it can take patience.

Until just several years ago, anyone could call himself or herself an organic grower and a good many did as the 1960s and 1970s spawned a heightened consumer awareness of farm-chemical problems and the advantages of organic foods. There were no national standards for nearly 30 years, as words like natural, organic, and chemical-free floated around like free radicals, meaning little or nothing on a package or farm sign. What few laws existed varied among states.

In 1990, Congress passed a law called the Organic Foods Protection Act (OFPA) to develop a national standard for organic production. The USDA was charged with developing regulations that would explain the law and oversee its implementation. The law provided for the establishment of a National Organic Standards Board to recommend what substances could and couldn't be used in an organic operation, and the board also helped the USDA write its regulations. Twelve years after the law was passed, the final rules were implemented in 2002.

Today, any farm selling more than $5,000 worth of agricultural products in one year and wanting to label them "organic" must submit to a rather strenuous certification process by a USDA-accredited agency. This cost in the neighborhood of $700 in 2006. Farms with sales less than $5,000 can call themselves organic if they

The National Organic Standards

The national organic standards don't simply specify that synthetic products can't be used on an organic farm; they cover record-keeping, how compost should be made, and much more. A copy is available free at the Appropriate Technology Transfer for Rural Areas (ATTRA) and the National Center for Appropriate Technology (NCAT) web site (attra.ncat.org/attra-pub/organcert.html). Both groups are umbrella organizations providing extensive research and information on organic farming. Or, you can get a copy of the standards from the U.S. Department of Agriculture (USDA) web site at www.ams.usda.gov/nop/ or by calling the USDA at (800) 346-9140.

Below is a summary from the USDA National Organic Program (NOP).

Organic crop production standards specify:

Land will have no prohibited substances applied to it for at least three years before the harvest of an organic crop. Use of genetic engineering, ionizing radiation, and sewage sludge is prohibited. Soil fertility and crop nutrients will be managed through tillage and cultivation practices, crop rotations, and cover crops, supplemented with animal and crop waste materials and allowed synthetic materials.

Preference will be given to the use of organic seeds and other planting stock.

Crop pests, weeds, and diseases will be controlled primarily through management practices including physical, mechanical, and biological controls. When these practices are not sufficient, a biological, botanical, or synthetic substance approved for use on the national list may be used.

Required information in the certification process for crop production includes: Land-use history documentation, field maps, crop-rotation plans, soil-improvement and pest-management plans, seed sources, material inputs (soil amendments, fertilizers, compost, manure, pest-control materials, or any other materials) used and planned for use, measures to maintain organic integrity (with regard to borders and buffers, application, planting and harvest equipment, post-harvest handling, and storage), planting, production, harvest and sales records, monitoring systems, and product labeling.

The organic livestock standards, which apply to animals used for meat, milk, eggs, and other animal products, specify:

Animals for slaughter must be raised under organic management from the last third of gestation, or no later than the second day of life for poultry.

Producers are required to give livestock agricultural feed products that are 100 percent organic, but may also provide allowed vitamin and mineral supplements.

Organically raised animals may not be given hormones to promote growth or antibiotics for any reason. Preventive management practices, including the use of vaccines, will be used to keep animals healthy.

Producers are prohibited from withholding treatment from a sick or injured animal; however, animals treated with a prohibited medication may not be sold as organic.

All organically raised animals must have access to the outdoors, including access to pasture for ruminants.

(Note: A civil penalty of up to $10,000 can be levied on any person who knowingly sells or labels as organic a product that is not produced and handled in accordance with the National Organic Program regulations.)

follow the standards, but they cannot display the USDA Organic Seal.

Most conventional farmers transition to organics gradually, one bit of acreage at a time, while others take what one grower calls the "suicide" route and do it all at once. That often occurs when a farmer begins to get interested in organics, begins reading up on it, and suddenly decides to just go for it.

Others prefer to go slow and prepare. The three years it takes to transition from conventional to organic may seem an eternity, so the more work you can do on the front end, the better.

Your first goal, again, is to study the standards and know what they mean. Come to know what organics is all about. Research takes time, but it costs little or nothing. Visit

Seeds NOP rules require that farmers use organic seed whenever possible in order to be certified. You can find organic seed at some seed houses.

Until recently, anyone could call their produce organic (or natural, which is devoid of meaning), but now produce called organic must be certified under national standards.

with organic farmers or those who are in transition. Send away for studies that have been done. Look at the list of resources, printed and otherwise, including those in the back of this book. Learn everything you can, and try to get the mindset of a land steward, and/or a soil steward, before going ahead. Then visualize what you want your farm to be 5 years hence, or 10. Set goals. In fact, set HIGH goals. If you set them, and visualize meeting them, you'll get there.

Don't stress about financial goals, however. These will take care of themselves. They're like grades in school: If you show up, pay attention, and study, grades take care of themselves.

Then come up with a transition plan. Figure out what's going to get you from here to there. Figure out what on the farm must change in terms of equipment, supplies, crop rotations, weed control, all of it. You might be surprised to find that many assets on the farm will also suit for organic production, or can be modified to serve. That old manure spreader rusting behind the tractor barn might be put back in service for spreading compost, for example. Drip-irrigation systems can be used for fertilizing organically as well as chemically. A green-house is the same way.

In setting up your transition plan, be sure to put it on paper. Break it down into the simplest parts, then expand on them in detail as needed. But keep it simple. The plan will change as you go along, but that's okay.

Begin laying the groundwork for your transition. Have your soils tested to see where you stand, and begin adding what you need. Look for sources of compost materials and manure. Be creative. Go not just to dairy farms, but to the extension service, racetracks, county fairs, any place where you'll find a good supply of organic manure, straw, wood chips, whatever you may need. Start with your own land first, however. Start compiling your compost materials immediately, and begin building up the soil with them. Unless your compost piles or windrows are moistened and turned regularly, it will take a full year or more for plant material to reach compost stage; longer for wood chips.

Remember, too, that in terms of sheer organic mass, relatively little leaves the farm as produce. Most organic matter remains on-farm as leaves, stalks, and root mass. If that and green-manure crops are turned under with a shallow cultivation system, your soil will be headed in the right direction from that alone.

Make lists of your current inputs and their costs, along with projected new inputs and their costs. Look for sources of certified organic fertilizers, wet and dry, and determine how much you'll need to get over the hump between chemical and organic. You may be surprised to find some real savings here, both short-term and down the road. Again, ask other farmers what they did, and look for help from the extension service. Most extension agents are deeply embedded in conventional practices, but states are slowly moving into the organic realm as pressures build in that direction.

If your land is dependent on chemical fertilizers, pesticides, and herbicides, their sudden removal may result in yield drops, so prepare for this in advance by cultivating, cover-cropping, crop rotation, and building your soil to reduce weed pressure and improve crop health. However, the USDA reports that many farmers suffer no adverse effects during transition. Yields often remain high (and in rare instances actually increase) and input costs often decrease with the adoption of organic methods.

Remember, you can certify your farm in sections; it needn't be all or nothing. If you have hayfields, for example, that haven't been chemically sprayed or fertilized for years, and meet the Organic Standards guidelines, then apply for certification on those acres immediately. Or if you have fields that have been relatively weed-free, begin by cropping those organically.

Consider the following as well:

To minimize disease, select fields with well-drained soils and good moisture-holding capacity that have been well rotated.

Choose crops that are relatively free of pest threats, such as cole crops in the fall and root crops summer and fall.

Sow in fields proven to have the best fertility and pH.

Use a vigorous-growing cover crop, such as winter rye, that can be sown in the fall, then rolled and crushed in spring for a natural mulch that will suppress weeds and hold soil moisture while a spring-sown crop matures.

Till as little as possible to reduce erosion and the loss of soil organic matter.

As the change progresses, you'll notice the soil becomes gradually spongier and less compact, less subject to crusting, more erosion-resistant, and more drought-resistant. Beneficial insect and bird populations will begin to return. You won't notice the changes right away, but one day you'll step back and remember how it once was and how it is now. More weeds may be present, but their effects won't be great because your crops can hold their own. On the financial side, you'll find input costs declining noticeably as you go along.

Microbial activity in the soil will gradually improve, although you won't see this, only its effects. The severity of any change varies greatly with climate, soil, cropping history, types and amounts of chemicals used, transition strategy, and other factors.

But all can be managed if you break the process into pieces.

Going into certification, you'll want to develop a comprehensive record-keeping system, along with a written description of your intended organic practices. Record-keeping and organization are an important part of organic certification, and they aren't difficult. They'll seem time-consuming in the beginning, and they are, but as you go along it all becomes routine.

Follow the national organic standards closely as you go along, and the journey will be not only healthier for you and your customers, it will actually be fun.

Before you know it, certification will be in place and you can begin to reap the rewards. In an article for The Rodale Institute's NewFarm website (www.newfarm.org), one farmer reported the results of his certification this way:

"Using organic management practices, I have minimized my input costs, increased my market opportunities, and, most importantly, increased my control over prices. For example, in 2001, my organic corn and soybeans yields met about 90% of my conventional yields. In marketing this harvest, I delivered conventional corn to the mill for $2.10 a bushel, while an organic processor picked up my organic corn at the farm and paid $4.70 per bushel. At the same time, I delivered conventional soybeans for $3.80 per bushel, while my organic livestock feed beans were picked up at my farm for $10.50 a bushel and my organic food-grade beans were picked up for $15 a bushel."

That works for me.

Chapter 12

TO MARKET, TO MARKET

Writing out the day's tomato order which just arrived by cell phone.

Two rules govern all there is to know about successful organic farming:

Quality is everything.

Marketing is everything.

Follow those two rules and you can't fail.

Most organic growers already know how to grow a superior product. That's why they grow organically in the first place. The area that can be tough is marketing. We have too much to do in the field, and if we were outgoing *marketing* types we probably wouldn't be farmers in the first place. Still, this one word—marketing—can help make or break an operation.

The good news is that you have an absolutely superior product to sell, and your attitude should be that of announcing from the rooftop that yours is the BEST.

Make checks payable to PEACEFUL VALLEY FARM

None better. It's more nutritious, it's more delicious, it's fresher, it's reasonably priced, and it's being offered for sale from a member of the community. The word community is a big one, maybe even as big as organic, because it emphasizes that you are one of them. A neighbor. Somebody who cares. And that's precious.

If you're selling wholesale to Wild Oats, for example, the produce manager will know many delivery people but very few growers. Your face and name will be one of those few. If he needs four more boxes of tomatoes by Tuesday, he'll call you and know that you personally will be there.

That's important in many ways, and one of them is rather subtle. It's that people like to know where their food comes from. Food is life, and we've found that people not only want to visit our farm stand, but want to visit our soil. They want to see *their* corn growing. They'd even enjoy a chance to pick it. That closeness, that connection may not be evident with produce managers, but I'm willing to bet it's there.

WHOLESALE

Wholesale selling for wheat and soybeans is one thing. For a small farm it's quite another, because so many other options are open. Wholesale can be a good outlet if you have enough to make deliveries worthwhile, but the price isn't nearly what you can get in direct-to-consumer sales. And with wholesale, your produce is anonymous; it's just right there with the other apples and tomatoes, with nothing to say it's yours.

Marketing in this case means getting your name on your stuff, so word can get around that yours is the best and here's where you can get it. Eliot Coleman has beautiful shallow boxes with the name of his farm burned into them, and these are what he uses to display his produce at local outlets.

Marketing means getting your name on your stuff, so word can get around that yours is the best and here's where you can get it.

Make your sign handsome, large enough to read from a distance, and facing up and down the road.

If you have the land and time, however, marketing through local organic food stores is another good source of revenue.

We set up displays of baskets and boxes with a large, laminated sign proclaiming "Plum Hill Farm Organics." This went into several stores, including a fish market that had never sold fresh corn and tomatoes until I suggested it to them. He liked our food so much, and knew it helped draw business his way, so he paid us full retail fare for everything sold, so long as we took back whatever spoiled.

We kept all sales to within a few miles of the farm, but found that even that amounted to more running around than we had time for. Eventually we cut back to the farm stand and restaurants only.

If you have the land and time, however, marketing through local organic food stores is another good source of revenue. Both Wild Oats and Whole Foods buy at least some of their produce locally, and non-chain stores do as well. It's worth looking into. Stop by with a basket of produce, talk to the produce manager, and see if you can't make a deal.

A farm stand should be at least as attractive and tasteful as your produce.

FARM STANDS

I can't imagine anything friendlier than a roadside farm stand. It brings back dusty memories of a different time when life was slower, and more than one customer over the years has put it precisely in those terms. And it wasn't just the farm stand itself; it was the farm. A farm stand without a farm behind it is like a barn without animals.

If you're contemplating selling from a farm stand, there's much to consider.

Location

First, you need a busy road relatively near a population center. Our stand was within easy reach of two towns, located on one of two main roads between them, so it was ideal. If your place is off the main drag, then you might be able to post a directional sign at an intersection, but this is not as easy as it once was, however, what with sign codes and zoning ordinances. If you can post a sign, follow the highway department's example: Make it large to be seen a way off, and high off the ground. It might be white-painted wood in the shape of an arrow with the words Farm Stand, or (better yet) Organic Farm stand on it. Give the sign all the simplicity and appeal of your stand itself.

Where we live now in Tennessee, neighbors with something to sell know the town inspectors don't work on weekends, so on Fridays the signs go up, and on Mondays they come down. You might try that.

Try anything you can to get word of your stand out there and develop a reputation. We made a beautiful sign with plums on it to stand by the road, and that helped announce that we were open for business. Whatever sign you put up, just be sure it reflects on the quality of your food. Don't be sloppy. And remember that you actually need two signs because one sign facing the road won't be seen. A sign must be perpendicular to traffic, not parallel to it.

Structures

As for the stand itself, we built a post-and-beam structure that practically shouted "old-time goodness." It was too close to the road to meet codes, but the new building inspector assumed it had been there forever and never bothered us. Things work out.

The simplest structure would have a shed roof enclosed on three sides and open to the road. If that means facing the stand west, however, then reconsider. You can't imagine how quickly fresh flowers and vegetables can perish under a hot afternoon sun unless you've been there. A morning sun is better, and a stand facing north or south is best. If your stand absolutely must face west, then try opening the north and south ends to display produce there.

You'll also need room for parking for at least six vehicles; eight is better. Crusher-run gravel makes an ideal parking surface and requires no base. We learned quickly that many people pass a farm stand by

Honor System

We've employed the honor system to good effect. Label everything for price, make the scale handy to both you and a customer (we used an old hanging scale) and put out bags, a calculator, and change. Leave a sign saying "Help Yourself, we had to run" (or some such) and you're in business. Some called us crazy, but we never lost a dime that we're aware of, and customers often left a note saying sorry they missed us.

Purple basil we also sell by the bag, and add it to bottled vinegar as a novelty item for salads.

Cosmos and dill are two of my favorite bouquet cuttings.

One key to success was having our upper market gardens be "show" gardens, in that we grew herbs, flowers, and vegetables there for sale, but customers were also free to wander through and see what was cooking.

unless somebody else is parked there. That car draws a second, and these draw a third, and so on. This takes on a life of its own. Parking your own car there as a "seed" vehicle only works for those who don't know you. We tried that.

Operations

If you have a farm stand, somebody has to be there to run it. The moment you run in for lunch, two customers will stop. The answer can be to eat in shifts, or eat under an umbrella behind the stand as we used to do. Keep the whole family involved. We found manning the stand to be at least as rewarding as growing produce. We made friends, traded recipes, filled special orders, and offered tours and advice.

When running a farm stand, it's tempting to work in the field while keeping

one eye on the customer parking area, but I don't recommend that. Customers will feel guilty for taking you away from your work, and you'll end up resenting the intrusion. It breaks up my concentration to constantly be watching out for two things at once. It's best to have somebody run the stand full-time, or at the very least be working within a few paces of it.

At our place, one key to success was having our upper market gardens be "show" gardens, in that we grew herbs, flowers, and vegetables there for sale, but customers were also free to wander through and see what was cooking. This meant not a single weed could show, so we could work in these areas and tend the stand at the same time. It also meant we could pick custom flower orders without breaking a sweat.

Flowers and nursery plants made up a big part of our business, and we promoted both heavily. We'd put the flowers into individual vases for sale, and also in buckets to make up arrangements. These looked wonderful from the road, and helped

business enormously. In flower arrangements we'd often add sprigs of dill or basil, especially purple basil with its striking blossoms, to add fragrance to our arrangements. These were usually impulse buys, but they made a difference. If you do grow market flowers, think long and hard about supplying weddings. In fact, don't think about it at all. Don't do it.

You'll want to keep regular hours, and be sure those hours are posted. Don't try to work seven days a week. It isn't worth it. We'd take Mondays off because they were always the slowest, with traffic increasing all week and then peaking on Friday night and Saturday. Sunday began to ebb again. Other farmers have experienced a similar pattern.

We had a competing farm stand in Massachusetts that purchased virtually all of its produce off-site, including citrus, and while there are no laws governing this, I advise against it. It turns you into a sales clerk instead of a farmer, for one thing, and I don't believe that customer loyalty is great in that case. Growing organically makes a

Our friends, Hank and Cindy Delvin are branching out into Shiitake mushrooms, which are grown on special logs and harvested over a period of several years. Always look for something that nobody else is selling locally.

The author with a pumpkin load. We've found that large pumpkins do well as a pick-your-own, while smaller ones sell well at market.

difference, mind you, but once again, I believe people want to see where their food comes from if possible.

Our farm stand was small and seasonal, of course. It ran through until cold weather sent us, our customers, and most of our produce inside. We were grateful for that because there's much to do in the fall and winter with composting, soil-building, mulching, and tending the greenhouses.

PUBLICITY

Keep in mind the possibilities for publicizing your stand. By making our stand so lush and beautiful by the road, we attracted the notice of a local stringer for the *Boston Globe* who did a story on our farm. The story began by stating that "A visit to Plum Hill Farm is like stepping into a lush, living painting." The story came out on a Sunday, and we had more than 400 customers that

When running a CSA, it's always wise to cement the customer bond with an event such as a cook-out with a hayride around the farm. *Photo courtesy of Delvin farms*

day, buying everything that was alive or recently so. Many of these we kept as loyal customers in future years.

I also wrote a gardening column for the local weekly newspaper, which Janet (a graphic artist and painter) illustrated lavishly. In exchange, the newspaper gave us free advertising space, so we could announce what was being harvested that week. We also got permission from a local bank to park our old two-wheeled buggy near their drive-through window, advertising our farm stand. Don't be bashful. Think of anything you can do to promote yourself in a respectful manner.

THE FARMERS' MARKET

A farmers' market is the absolutely best way for a newcomer to organics or farming to market produce. It gives you a chance to build up a steady customer base at a comfortable pace, without the need for advertising or any major overhead. All vendors at a market are on equal footing, with an equal chance to attract customers. All you need is booth space, quality produce, a friendly smile, and good signage.

Our own Saturday market charges only about $30 for a small booth area, and accepts weekly payments (cheaper by the season, though) so newcomers needn't worry about much investment.

If you already have a market within easy reach, spend some time there. Talk to the vendors. Explain your plans. Find out what sells and for how much, and how it's displayed. Find out if organics get a price premium (they don't where I live). See where you'll fit in. Look for a niche that isn't being met. A thriving market can easily support more than a few vendors selling vegetables, but if you add something distinctive to your table, it could make a big difference.

In our case, we sold vegetables and flowers only for several years until Janet decided one summer to make pesto out of our surplus basil. She brought some in one Saturday morning in 8-ounce canning jars, offered samples on crackers, and it was gone within an hour. She brought more the following week, along with salsa, and it sold out. Then she began making pickles, relishes, jams, jellies, breads, and more, and it all sold out. We didn't have time to make more, especially the pesto, so we raised prices. Eventually the pesto was selling out to the tune of several hundred dollars a week at $7.95 for an 8-ounce jar.

We still sold produce, of course, but people came to our booth for the pesto and

Farmer's Market Survival Tips

Personal survival tips for those participating in a farmers' market include:

• Have plenty of change on hand. We bring $100 in fives and ones.

• Have twice as many T-shirt bags as you think you'll need.

• Make a good presentation with your tables and crates so they are neat and full-looking. If a bushel basket begins to empty, put the remainder into a half-bushel, and so on, so it remains looking full.

• Keep standing whenever possible, or seated on a high stool.

• Make eye contact, smile, say good morning, and be sincerely interested in those who stop. But don't ever "hawk" your wares. Let your produce speak for itself.

• Invest in a large banner proclaiming your farm name and the words "certified organic."

• An easel or notebook with pictures from your farm helps personalize you.

• Bring a scale, calculator or cash register, pen and paper, and recipes. A water spritzer helps keep your produce moist and fresh-looking. If electricity is available, and it should be for cash registers, bring a fan for hot days.

• Label prices clearly, either on a big board or on individual crates.

• Price fairly, but don't be afraid to price at a premium. Just don't undercut the market.

• Make sure your customers write down their e-mail addresses so you can send them weekly notices of what's fresh and what crops are coming in at the farm. This list will prove to be enormously valuable to keep your loyal customers.

Basil is not only easy to grow and harvest, but an easy sell at a farm stand or farmers' market. Include a recipe for pesto. Other herbs, such as dill, cilantro, and chives, also sell, but not like basil in our experience.

pickles, and then picked up corn and cabbages.

Your own niche might be fresh herbs, or houseplants, or a tub of spring water on ice. The possibilities are almost endless, within the bounds of market rules.

A NEW MARKET

If you don't have access to a farmers' market, consider setting one up. You'll have good company: The number of local farmers' markets in America more than doubled in 10 years from 1994 to 2004, from 1,755 to 3,706. Today that number is estimated to be close to 5,000 nationwide.

We were fortunate to be in on the ground floor in setting up our own farmers' market five years ago, so we have a pretty good idea what works and what doesn't.

The first step is to find somebody with organizational skills who wants to help you do the groundwork—don't try it alone. This means talking with local officials, business people, residents, and perhaps the extension service to get their support, and perhaps even sponsorship, in advance. Get word out that you're setting up a market and talk to potential farm vendors as well as those with a potential location.

The next step is to do some preliminary research on location. Look for the following.

• A central location within easy reach of customers, and reasonably easy reach for vendors. Customer convenience is primary.

• Plenty of easy parking. Shoppers demand this, and won't solicit a market where they have to walk long distances carrying a half-bushel of produce.

I consider smallish cabbages not only tastier, but easier to sell.

• Weather protection. A market need not be enclosed, but you need a cover.

• Safety. The location must be a safe place for crowds, especially children.

• Easy vendor access. This means vendors must be able to back up their trucks and unload easily.

• Restrooms, water, and electricity handy.

• Places for customers to sit. This isn't mandatory but it helps.

• Reasonable rent. The price of a location greatly impacts how many vendors you'll get and how happy they'll be. If possible, get a location for free so you can put vendor rent toward advertising and other costs.

Determine early on what type of vendors you want. Farmers are obviously needed, but you must decide whether the market should be limited only to locally grown produce. Or you can require that a certain percentage of goods sold must be either grown or made (in the case of craftspeople) by the vendor. The emphasis on local produce and crafts is a major attraction in drawing customers to a farmers' market, so I would be wary of allowing in too much merchandise that one could buy elsewhere. A farmers' market needs to be unique.

At this point you're ready to call a meeting of potential vendors. Put an ad in the appropriate newspapers and put up flyers at the local Tractor Supply, farmers' co-op, or other appropriate places. At the meeting you will be able to hash out vendor interests, locations, days of operations, and other concerns. Be sure to collect names, addresses, and phone numbers. The first meeting may be sort of a free-for-all opportunity to get to know each other, ask questions, and get the ball rolling. At a subsequent meeting you'll want to select a name for the market, set up a board of directors, seek out a market manager, and begin narrowing down choices as to location, days and hours of operation, bylaws, and market rules. The market manager may be a volunteer in the beginning, but most are paid and some are full-time in the larger markets. It's a big job, and you need just the right person to balance opposing interests.

The board or manager will deal with insurance and safety matters, along with state and local health codes regarding meat and poultry products, processed food, and value-added food items. For example, open samples of raw or processed foods may be prohibited in some areas, and the use of a certified kitchen is often required for canned food

such as pickles. What's acceptable at a church bazaar may not be, and probably won't be, in a farmers' market.

On market hours, it's wise to limit the market to one day a week at first, and Saturday is usually the best, especially the morning. Our market is open from 8 a.m. to noon, and all vendors are expected to be fully set up by 8. Most vendors are local, but we drive 40 minutes and other vendors drive two hours to get there. More than one market day, or an all-day market on Saturday, may take too much time out of the growers' week, but that's a decision to be made by every market. Getting up at 4:30 a.m. to pack, drive, unpack, and be ready by 8 makes for a long day.

Initially we had a Tuesday market also, but it was poorly attended and we ended it. The best idea is to start with a four-hour market and see how it grows. Don't expect a lot of activity the first year. It took two years for our new market to develop a good customer base, even with newspaper ads and local TV coverage. Once you're up and running, don't neglect to inform local news outlets of what you're doing. It's even wise to drop off a box of market samples at the newsroom now and then until you get noticed.

Our market chips in the pay for the services of a local bluegrass band, and that always helps to draw a crowd. Barring that, it always helps to locate the market where

people normally gather, such as near the post office, soccer field, or another community-related locale.

COMMUNITY-SUPPORTED AGRICULTURE (CSA)

The advent of CSAs about 20 years ago is about the best thing that's ever happened to small farmers—and to communities, for that matter.

The way it works is simple: A farmer sells "shares" of the farm harvest in advance of the growing season for a set price. In return, members are guaranteed a share of fresh produce every week or two through the season. The arrangement guarantees the farmer a market for produce, and gives the shareholder an easy, local, fairly priced source of fruits and vegetables (or meat, or poultry, or all of the above).

For example, a full share might cost $700 for a half-bushel of produce every week from May through October. A half share at $400 would buy a half-bushel of produce every other week. This is what one of our neighbors charges for certified organic produce, and if the price by weight were averaged out over the season, it comes in at less than what one would pay at the supermarket. Shareholders either come to the farm for their produce, or come to one or more pickup sites during the week. My neighbor's 76-acre farm has only about 40 acres in cultivation at one time, and

Roma tomatoes, already graded, being washed on a sloping tray over a long water trough. Always let customers know your produce has been thoroughly washed. This is added value.

A successful CSA or farm stand grows 30 or 40 vegetables with a succession planting system that makes a wide variety of items available all season long.

supplies 340 CSA members—most of them full shares—as well as several stores and the farmers' market. They also have anywhere from three to six helpers at any one time when the 28-week season is in full swing. In other cases, shareholders may actually volunteer at the farm to help the grower. It all depends on how you want to set it up.

Always start small. You don't want more shareholders than you can supply with produce all season long. That means a lot of organization on your part to be sure that you can supply a wide variety of fruit and vegetables week after week throughout the growing season. A successful CSA or farm stand grows 30 or 40 vegetables with a succession planting system that makes a wide variety of items available all season long. If you do this, grow everything the other farmers do, and also grow what they don't. We didn't sell a lot of herbs, for example, but our customers love the fact that we had them.

A typical spring offering might include strawberries, broccoli, cabbage, green onions, kale, collard greens, turnip greens, radishes, and lettuces. Summer brings carrots, blueberries, tomatoes, bell peppers, a variety of hot peppers, peas, squash, zucchini, eggplant, kale, melons, and Irish potatoes. Fall crops include turnips, kale, collards, beets, tomatoes, sweet potatoes, cabbages, turnip greens, peas, sweet and hot peppers. Greens should be available almost year-round.

Always make sure the box overflows and has nice presentation on top. That means mixing colors, textures, and the like. Put potatoes on the bottom.

To get started, talk to other CSA farmers. The extension service will help with names. You'll need some good farming experience, either from years of doing it, or by apprenticing at an organic farm for a season. Plant your own market gardens, read, ask, and practice. And be

Mustard greens make an excellent addition to our mesclun mix.

prepared to work hard. If the sun is up, you will be, and you may well be working long after it's down.

The first year is often the worst, because it may be fraught with doubt on your part. It helps to know that others have done it and you can, too. The idea of growing produce good enough to sell is a big leap when you first do it, so remember you not only can do that, you can grow the best around. It helps to have family members who can chip in, and once you're up and running, you might find interns willing to help for little or no money, but for the experience of learning organics.

At Plum Hill Farm, we had one paid helper and several unpaid interns to help out. We always served them a delicious lunch and made sure they had a hand in decision making. We all worked together, and worked at a fast pace but took all the breaks we needed. I remember one day when one of our interns commented that farming was a lot of manual labor, to which I replied that if working with the soil was manual labor, then making love is manual labor, and playing a violin is manual labor. It's all in your perspective.

Set up a core group of members with whom you can consult regularly. These people might help out with newsletters, deliveries, or other tasks when the work overwhelms you.

Be sure you have an e-mail newsletter to keep shareholders informed as to what they can expect each week.

Have a cushion of produce. Once the CSA orders are filled for the week, the excess may be sold at a farmers' market or at a specialty store. But always have a cushion.

Consider restaurants. The good ones want only the freshest ingredients, and if they can buy locally they will. Select a few and bring them a choice of your produce. We had wonderful success with mesclun greens because they looked so much more exciting than a bed of iceberg.

Develop a network of other farmers who might be able to help out if one of your crops fails, or to supply your needs for a crop that you can't or won't grow, such as sweet corn. One CSA neighbor of ours relies on an Amish farm for sweet corn, and on another organic grower for okra, because both are crops he doesn't enjoy growing (or harvesting, in the case of okra).

Spread word of your CSA by posting fliers at appropriate places, placing an ad in the local newspaper, or getting free editorial coverage if you happen to be the first such CSA in the area. Many, no *most*, people still aren't aware what the initials mean, and newspapers are always looking for feel-good stories.

And few things are more feel-good than an organic farm.

Americans spend a smaller share of income on food than do people in any other nation.

Chapter 13

VEGETABLE GROWING

I would feel more optimistic about a bright future for Man if he spent less time proving that he can outwit Nature and more time tasting her sweetness and respecting her seniority."

—E. B. White

One thing we've discovered over the years is that growers seldom agree on how to grow certain crops, and the same holds true for seed catalogs and books. Compare any two catalogs as to how far apart to plant any vegetable and you'll see what I mean. I've also found that crop information for home gardens can be wildly at odds with strategies needed to sell on the market.

To remedy this, I've put together the best advice I could glean from 50 years of growing produce and 20 years of selling it.

In general, you want to develop a reputation for growing a crop that not only looks delicious but tastes that way, outshining any your customers could find elsewhere. Then you must do this every year, consistently. Make one slip, sell one bad tomato, and they'll remember. Grading, in fact, can be just as important as growing.

Try to be wary of new or unusual vegetables, such as purple carrots and white beets. We've found that most customers are very conservative when it comes to their food.

Make sure your produce is carefully washed, and just as important, make sure your customers *know* it's been washed. That's a value-added thing.

Of course, the first trick is to grow the absolute best in whatever manner works for you.

ARTICHOKES

Globe artichokes are not a wildly popular vegetable, and most would not expect a grower outside of central California to offer them. That's why you should. They're finicky but not altogether difficult to grow from seed for harvest the first year. They're rich in vitamin C and low in calories. They're also a heavy nitrogen feeder.

The artichoke is actually a thistle and a member of the sunflower family whose mature head is an unopened flower. The tender bases of the petals and the fleshy heart to which the petals are connected are the edible portions. In warm areas, the artichoke is a spreading perennial, but is now available as an annual, with six to eight heads on each 3-foot-tall plant.

Start seed eight weeks before last frost in 2-inch or larger cells, where they may be grown cool (as low as 60°F days and 50°F nights). To induce early budding, either cool seedlings for 10 days in a fridge, or set out transplants when they can get more than a week of temperatures around 50°F or lower (but protect from frost). This fools the plant (normally a perennial) into thinking it has experienced winter. Set out plants 2 to 3 feet apart.

Artichoke roots must be kept cool, and they're best grown quickly, so keep the soil moist and mulch heavily against hot weather. Intense sunlight and high temperatures may cause the plant to wilt, even when

---

Tip

Don't spend too much time or land on artichokes until you find that your soil and climate are suitable for them. This is something to experiment with because the globes make a nice presentation for sale, and generally won't be expected.

Asparagus is known as a spring crop, but can be harvested all season

soil is moist, but it will recuperate later. Intense sun may also cause the plant to grow quickly and put out small buds, but when the weather cools the plant will once again produce nice-sized globes for six weeks or longer from midsummer into fall, depending on location. They prefer a slightly acid soil of about 6.0.

ASPARAGUS

Asparagus is a perennial plant that will produce for 20 years or more, and can be grown from crowns or seed (and the seed is far cheaper if you have a heated green-house). Asparagus supplies more folic acid than any vegetable, and leads most vegetables in a wide array of nutrients including potassium, thiamine, and vitamin B6. It also has only 4 calories per spear.

Asparagus is an investment, so prepare the beds well with as much compost as you can spare and a soil pH of 7.0 or higher. Space plants 12 to 18 inches apart in rows 3 to 5 feet apart. In a hole 6 to 8 inches deep, set the plants (or crowns) flat on the bottom in early to mid-spring, then fill in with compost gradually as the plants grow over several weeks. Let the ferns grow without harvesting stalks

Tip

1. Most people harvest asparagus only in spring, but it's just as easy to harvest later when nobody else is selling. Just mark the rows for delayed harvest, let the plants go to fern, then snap off the ferns and harvest the spears that come up. Keep harvesting spears as you would in spring, then let the plants go to fern again. Presumably the plants don't care whether the fern feeds the roots, early or late. I've experimented with picking spears through late summer, but always try to let them end the season as ferns. I don't know why; just a feeling.

2. To prevent crown and root diseases, spread two pounds of sodium chloride rock salt (NaCl) per 100 feet of row every spring before the tips appear, or in late summer. This also seems to improve growth. Pickling salt also works, but don't use table salt or rock salt made of calcium chloride (CaCl).

using a knife, which isn't necessary.

When planting from seed, sow seeds at least 12 weeks before transplanting out after danger of frost. Asparagus is slow to germinate (about 30 days), therefore I sow all seeds in 32-cell flats to avoid transplanting. When the plants are about 12 inches tall they're ready to move outside. Consider moving them to a nursery bed the first year (planted 3 inches deep) to separate male from female plants. A bed of all male plants can out-produce a mixed male/female bed by up to 30 percent because of the energy females put into seeds. Females also don't live as long, either. To tell the two apart, wait until your plants produce flowers, then check them with a magnifying glass. The flowers of both are small, bell-shaped, and whitish green, but male flowers are larger, longer, and more conspicuous than female flowers. Eliminate the females and sow the males in a permanent bed the following spring.

Cooking: The best, no I think the *only*, way to cook fresh asparagus is to sauté in a covered skillet with the bottom just covered in salted water until the stalks turn a bright green without getting soft.

BEANS

Beans are one of my favorite crops because they are so trouble-free, so delicious, and so good for the soil. They're a snap to plant and look so good at market—there's something soulful about a basket full of fresh snap beans. They're also an excellent source of fiber, protein, calcium, and iron.

They can be time-consuming to pick by hand, but I rather enjoy going down the rows seeing how fast I can fill a container. A mechanical picker, justifiable only on a large-scale operation, can result in a lot of low-priced beans (less than $1 a pound at our market), but I've found that most customers will pay the higher price for organics. What's more, we're able to stagger our plantings to get beans long after the mechanical pickers, who in any case didn't make enough to stay in our farmers' market for long.

The thin, tasty French "fillet" beans, which became increasingly popular about 10 years ago, are indeed delicious and fine to look at, but to take full advantage of their gourmet price tag, one must pick them small and at least every other day. I find them harder to hand-pick than other varieties

until the second year, and then pick only the thickest stalks. Ferns feed the roots, so let them die back naturally in the fall.

Mulch heavily with straw or leaves around the plants annually to keep out any perennial weeds, and sow a leguminous cover crop between the rows. Till lightly in early spring before the spears appear, or turn under debris in late fall and mulch heavily with leaves. Add several inches of rich soil or compost to the rows each year, as crowns tend to work upward and produce a thinner crop over time.

Spears may be harvested for four weeks in the third year, and for eight weeks after that. Harvest stalks by snapping them off by hand just below the soil surface. This is faster than

Because we have numerous rails too short for fencing, we use them for pole beans, supported by sissal twine wound around the uprights.

such as (my favorite) 'Blue Lake,' and they didn't sell well enough to justify the effort. My experience is that most consumers are quite conservative when it comes to food, and while they'll pay extra for top-quality produce, it had better be of the highest quality, and not many of our customers thought the fillet beans made the grade.

We grow both the pole and bush varieties. We sow bush beans in a wide row, closely spaced to no more than 2 to 3 inches apart (contrary to most conventional wisdom). The thick bed drowns out all weeds, reduces moisture loss, leaves less open soil in a field for weeds, and makes far more efficient use of the land than one-bean rows. Picking is a bit more difficult in a wide row, but that's more than offset by the advantages.

Pole beans take longer to mature than their low-country cousins (10 weeks or so), but I find them more flavorful, longer-bearing, and more productive per square foot of land. We charge a small premium for them and get it. For supports, I use teepees made of cedar fence rails where one end has rotted off and is therefore no good for fencing. I wind sisal twine around the teepees and plant thickly around the bases. There are many other ways of growing pole beans, but I happen to have the rails in abundance, and they look good standing there like old wigwam frames. We also grow pole varieties along a stock fence on the edge of the property.

We get Mexican bean beetles occasion-

A mix of orange and red beets makes a nice display. *Photo courtesy of Fresh Harvest Co-op*

ally, but they tend to be an end-season crop—much like squash bugs—and don't do much damage to the beans so we ignore them.

On crop timing, I've never found it worthwhile to rush the season with beans. Organic growers can't use seed treated with fungicides such as Captan or other rot-preventatives, so it's best just to sit tight and wait until the daytime soil temperature reaches 60 degrees F, or a week or two after last frost. Then replant every few weeks throughout the season. I've read where some try to get a head start with transplant beans, but I don't think it's worth the effort and space in the greenhouse. Go with Mother Nature's timing on this one.

Harvested beans don't need to be chilled from field heat, but they do like high humidity and good air circulation. Cooling won't hurt them.

We only grow snap beans for sale, but grow shell beans for our own use. It simply takes too much time and effort to grow

either shell beans or lima beans for market. Wax beans we've found don't have much of a following (except with me). Flat-podded Romano pole beans have an excellent flavor, but we didn't find them as heavily-bearing as other varieties.

BEETS

Beets are one of my favorite vegetables, whether as greens in a salad, as baby beets bundled for sale or eating, or as mature beets in midsummer and afterward. We mulch some late ripeners and leave them in the ground for picking all winter under the snow or frost. They keep well so long as they don't freeze or get dry.

Beet greens contain more iron and minerals than spinach, while the roots are rich in potassium, fiber, and other vitamins. Don't overcook your beets as they lose vitamin C in the process.

One beet "seed" is actually a fruit containing a cluster of up to eight seeds.

This requires early thinning, some say, but I've found that one seedling generally dominates the rest and takes over. Quick growth with plenty of moisture is the key to tender beets.

Transplanting is possible, but I prefer to direct-seed these and all other root vegetables in loose, loamy soil. We grow in a 12-inch wide band, 1/2 inch deep and about 2 to 3 inches apart after the soil has warmed in spring but is still quite moist. Then sow in two-week intervals until eight weeks prior to expected heavy frost, although we've found that beets sown in midsummer heat don't do that well. We don't thin the bed per se, but instead thin as we harvest greens and baby beets. The greens we mix into our mesclun mix, with the 'Bull's Blood' variety being our favorite for both color and texture.

We seldom have any problem with pests. However, a soil boron deficiency can cause a lack of growth or browning of the beet hearts, especially in dry weather. A foliar spray of liquid seaweed helps, but it's best to avoid the problem by applying rock phos-

Transplants for cold weather crops, such as broccoli, can be set out in midsummer to ripen with a light frost or two to improve flavor.

phate or boron to the soil ahead of time.

Chill beets and greens quickly after harvest to remove field heat. I've had little luck selling beets at market with the greens attached, since they wilt much quicker than, say, carrot tops. Chilling and spraying may help keep the tops fresh-looking, but we prefer to sell the beets alone.

BROCCOLI

Broccoli is one of my favorite crops, not only for taste but for the bountiful look of its plants. It's a rich source of vitamins C and A, freezes well, and tastes good cooked or raw. And, it's become increasingly popular in recent years after medical research determined that broccoli (and other members of the brassica family) contain a wide range of cancer-fighting compounds, some of which are largely responsible for the taste and smell of these vegetables. Scientists at John Hopkins University have determined that broccoli sprouts have at least 20 times more of one such compound than a mature head.

Broccoli is best grown in cool weather, and the plants seem to do particularly well in soil with high nitrogen content, either from a previous crop of legumes or leguminous green manure. Broccoli also loves steady moisture for quick growth, so provide it with irrigation or soil with a high moisture-holding capacity (rich in organic matter).

We sow a spring and a fall crop, and the fall-maturing crop is generally the tastiest because it matures better and faster in cooler weather than the spring crop. We have few problems with pests or disease, outside of a regular infestation of imported cabbageworms, which are easily dispatched with Bt. Fall crops have almost no pest problems.

I grow only from transplants in this case. Spring seedlings should be transplanted when they're about 5 inches tall with four or five leaves, and when evening temperatures are consistently above freezing and preferably close to 50°F. Transplants that are set out too early in terms of cold weather, or too late in terms of seedling size, can produce small, immature heads called buttons.

Put the plants about 1/2 inch deeper in the soil than they were in the cell for good early root development.

Spacing recommendations for broccoli transplants tend to be all over the map, but I've had the best success with plants spaced 18 inches apart, two rows to a bed, and staggered such that four plants form a parallelogram when seen from above, not a square. The beds themselves are 30 inches apart. When planting in rows, I'd give 6 to 8 inches more distance between plants.

Broccoli and the other cole crops can be undersown when grown with wider spacing, but we've found that our close spacing in a bed precludes any serious weed problem because the soil is so shaded.

Harvest spring broccoli when the buds are tight but swollen, with no yellow showing, and cut to get as much stalk as possible. This part of the plant is just as tasty and nutritious as the head itself, and I hate waste. Harvest in the morning to avoid wilt, and cool the heads immediately. One or more harvests may be gathered from the side shoots, or florets, which are the plant's last-ditch attempt to flower and produce seed with the main head gone. We charge a premium for these shoots.

If growing a fall crop, sow a fall crop so that it matures a week or so before the first frost date. Frost slows its growth almost to a stop, but doesn't hurt the head a bit. I've seen heads turn purple and freeze, only to thaw out and end up on the table tasty as ever. You won't get much in the way of side shoots on fall broccoli.

BRUSSELS SPROUTS

Brussels sprouts are similar to broccoli and other brassicas in their healthfulness and culture, but these little cabbages are so absolutely superior in taste after a few good frosts that I don't even consider an early crop. Many consider them a delicacy, while others . . . well . . . don't. Sprouts generally require more than 100 days to maturity from transplant, so we sow to have a crop ready just prior to first frost. Nothing we know of can stand the cold like Brussels sprouts. They can be picked for Christmas dinner in most areas, and we can usually harvest through winter in our zone, 6b.

As with other brassicas, sprouts need fertile soil and good irrigation. I grow only transplants in 32-cell flats to transplant in six weeks. Plants are spaced in beds the same as broccoli, 18 inches apart with two rows staggered. Because these plants are set out when harsh summer weather is in full swing, to avoid transplant shock I give each transplant a drink of compost tea, fish

Tip

Some people have bad memories of having eaten bitter-tasting sprouts as a child. This is because they weren't cooked properly. Cook sprouts of similar size together, only long enough so that they keep their bright green color. Braising (in a little water in a tightly covered pot) works well. Sprouts cooked until they're dark will have lost not only flavor but nutritional value.

Tip

When harvesting, cut the head off above the outer leaves, not at ground level. This often results in smaller cabbages forming around the cut, growing to 4 inches or so.

Cabbage that's overcooked gives off unpleasant odors and loses its nutritional value rapidly. Cook briefly, until barely tender, and never in aluminum pans (which exacerbate the odor).

emulsion, or (better yet, in my opinion) a diluted shot of the liquid from our worm composting bins.

Harvesting usually begins about three months after setting the plants. Pick when the sprouts are firm, starting with the lowest on the plant first lest they start to open and turn yellow. A sprout's delicate flavor is lost at that stage. When picking, break the leaf below the sprout and push the sprout off the stalk. Upper sprouts continue to form and swell in the axel of each leaf as the lower sprouts are harvested.

Picked sprouts may be packed in quart berry boxes, or may be "topped" in early fall to leave you with a stalk full of mature sprouts ready to market. To do this, pinch off the growing point at the plant's tip (a cluster of small leaves) when the lower sprouts are no more than 3/4 inch around. In a month or so, all sprouts on the stem will mature to about the same size. These stems are quite attractive and easy to market. Moreover, with a stalk weighing two pounds or more, at $2 a pound (or more), what more could you want in an easy market crop?

CABBAGE (Brassica oleracea)

Ah, cabbage! What an underrated vegetable. I think this is because the only cabbages you find at the store are about the size of St. Louis. Bring one of those babies home, and all of a sudden the whole family decides to spend the night at friends' houses. Cabbages are one place where bigger isn't better.

The smaller ones have a more delicate flavor, so we grow only small to medium-sized heads. Sow one crop for early and

midseason harvest about 60 to 70 days after setting, and a second crop of midseason and storage varieties for fall harvest. The first crop grows quickly and has a small window of opportunity for harvest. Late cabbages mature more slowly, and can remain in the field much longer, allowing flexibility in the harvest.

Nutrients, pests, and cultural information are about the same as for broccoli and the other brassicas. Grow the early cabbage fast, fertile, and moist. Cabbages at our place get a little more favorable attention from imported cabbageworms than broccoli does, and we have to be careful about heads cracking, but the rest is about the same.

Cracking, or splitting, is caused by the same factors that cause tomatoes to split: A long dry spell followed by heavy rain or irrigation causes the fruit to literally outgrow itself and break apart. There is not much you can do except to be prompt with the first harvest.

Again, we grow only from transplants. Seeds like warm temperatures to germinate (at least 70°F), but don't mind cool temperatures after that. Come to think of it, the same is true of most everything in our greenhouses except tomatoes, peppers, basil, and a few other heat-lovers.

We sow all late-season brassicas outside in flats, and find that some varieties germinate in 48 hours. *This* hurries a crop along.

Purple cabbage (they call it red, but it's purple) is a highly attractive market crop and we sell it at a slight premium. It's excellent in slaws or as a side dish. As with green

We mixed purple cabbages with other vegetables and flowers in our display gardens, just for fun.

Orange sells best, but carrots come in a host of colors and varieties.

cabbage, we and our customers far prefer relatively small heads, no more than about 5 inches in diameter.

CHINESE CABBAGE
(Brassica rapa, Pekinensis group)

This tasty Asian specialty combines the fine crisp texture of lettuce with the juicy taste of mature cabbage, but with lighter green leaves. It comes in both barrel-shaped-heading types and the looser lettuce types, much like romaine lettuce. Both are used fresh in salads or sandwiches, or steamed, boiled, or stir-fried.

Chinese cabbage tolerates hot weather, and can be grown for both summer and fall harvests. For summer cabbages, sow in the greenhouse four to five weeks before last frost, or direct-seed after last frost. Space the plants 12 to 18 inches apart in rows at least 18 inches apart, or in 2- to 3-foot-wide beds with similar spacing. Choose only bolt-resistant varieties. Fall cabbages may be started from seed any time after late May, depending on region and variety. Chinese cabbage seems happiest maturing in the shortening days of late summer and fall.

Flea beetles are a common pest on early cabbages, less so on the late harvest. Chinese cabbages can be stored for several months at near-freezing temperatures and high humidity. I've heard they root-cellar well, wrapped in newspaper, and while I haven't tried it, regular cabbage is a good keeper this way, so I've no reason to think its oriental cousins wouldn't be.

PAK CHOI
(Brassica rapa, Chinensis group)

These attractive, nonheading cabbages also go by the name pak choy, bok choy, and others. They're all the same vegetable, and it, like all brassicas, is a member of the mustard family. Its distinctive dark green leaves with thick, light, succulent bases have a distinctively sweet, mild flavor used in Asian cooking for nearly 2,000 years. Pak choi can be steamed, stir-fried, or added to soups, stews, and salads. It will keep for several weeks if chilled at near-freezing temperatures and high humidity.

It's a cool-weather crop, preferring temperatures below 80°F and a quick moist growth period, maturing 40 or more days after direct-seeding, depending on variety and weather. (It's best grown from seed to prevent transplant shock and premature bolting, but others have done fine as transplants.)

Pak choi can be grown in spring for an early crop, or sown in late summer for a fall crop. Either way, the plant seems to gather lots of attention on our harvest table. Be sure to include a recipe for preparing it, however, since it's still new to many customers.

If you sow in spring, be sure to select only slow-bolting varieties. Harvest in the morning or in cool weather to prevent wilting.

CARROTS

Carrots are America's fifth-most-favorite vegetable, according to the USDA, but you'll rarely find anyone describing this root as flavorful. That's until you've selected varieties for flavor alone and grown them in rich, loose organic soil and harvested after they develop a bright orange color. That's when the flavor develops fully.

For *truly* sweet carrots, unlike any you've eaten, mulch them under a deep bed of straw or leaves to be picked and eaten all winter. The cold apparently converts starch to sugar, and is reason enough to keep the farm stand or CSA open a little longer in the year. Moreover, a single carrot root contains a full day's supply of vitamin A, along with vitamins B, C, D, E, and K.

As with other root vegetables, I plant carrots in a wide bed with five or six rows per bed. This not only allows more efficient use of land, but also allows carrot tops to shade out some weeds, and provides easy access from both sides for hand weeding and cultivating. It's virtually impossible to grow good carrots without at least some handwork. It's also effective to plant in rows several inches wide and about 24 inches apart, if you choose.

Once the bed is prepared and seeded, it's wise to scorch it with a propane flamer a day or two before carrot emergence. Again, this can be determined by placing a pane of glass over the bed to warm one area earlier than the rest. When seedlings appear under the glass it's time to flame. Carrots may crack, put out side roots, or have a poor taste in the presence of too much nitrogen, so plant them at the end of your rotation, after corn for example. Also, make sure your beds for carrots and other root vegetables are covered all winter in straw or leaves, as this virtually eliminates weed problems the following spring.

To develop good root size and shape, carrots need to grow rapidly and without restriction. Prepare a fine seedbed ahead of time with no germinating weeds. If the soil is compacted then deep ripping with a chisel plow may be needed—or loosening with a broad fork—to a depth of 12 inches. Some say carrots benefit from an even deeper loosening, but this may not be practical. (Cultivation techniques that avoid soil compaction, such as raised beds, are the real answer in the long run.)

> ### Tip
> If you're looking to lose weight, eat a lot of celery. You'll use up more calories chewing than you'll gain from swallowing.

Celery loves mucky soil and grows slowly, but is worth the effort.

A popular bumper sticker at many farmer's markets.

We sell carrots bunched with the tops on because they look so good, misting them regularly. Or cut much of the tops off to prevent moisture loss.

CELERY

Celery has something of a reputation for being difficult to grow, so most gardeners and farmers I know shy away from it. That's too bad, because the taste difference between organic celery and its chemical cousin is so dramatic that I almost consider the two as distinctly different vegetables.

Celery is a heavy-feeding, long-season crop, and to grow it you need highly fertile soil, a constant supply of moisture, and relatively cool temperatures with a pH of 5.2 to 6.6. The need for steady moisture cannot be overemphasized: Celery is 94 percent water. Wet, boggy places make an ideal site for growing celery. If you have sandy loam soil with no drip irrigation, then mulch your celery heavily to keep its roots moist.

Grow only from transplants, because the seeds take two or three weeks to germinate. They transplant well, however. Sow seeds about 10 weeks before setting out, when the weather is warm and settled (as you would for peppers or basil). Celery will bolt if it's hardened off by lowering temperatures. It doesn't like to be below 55°F for any length of time.

Set out transplants 8 to 10 inches apart in wide beds, or in rows 18 inches apart or more. Wider spacing can result in the stalks opening up more than you want for appearance's sake. Celery is often blanched as it grows to keep the stalks from getting bitter, but this results in nutrient loss and I'm not sure it's necessary. Some do it, some don't, but I've had delicious unblanched celery and hope to continue that. Store celery as you would cabbage. It will keep for 2 to 3 months.

Celery isn't subject to many pests, but may occasionally see damage from nematodes, aphids, and the carrot fly.

CHARD

Swiss chard is one of the easiest greens to grow, and can be harvested almost all season long from a single sowing, although we prefer two or three for baby greens. We love it in salads, steamed, sautéed, or stir-fried briskly. It's colorful, tasty, and one cup provides an excellent supply of vitamins K, A, C, E, magnesium, manganese, potassium, iron, and dietary fiber.

Chard is actually a member of the beet family that's been bred for its greens instead of its roots. It can be grown as transplants sown four to six weeks before setting out after frosts have settled down, but I prefer direct-seeding since the seed germinates readily. In rows 12 to 24 inches apart space seeds 2 to 3 inches apart, or plant in a wide row with seedlings thinned to 6 to 8 inches apart. A crop may be sown every two weeks for a continuous supply of young greens, but I never have time. Besides, the mature greens are tasty enough as it is.

Begin harvesting about 60 days after sowing, taking the outer leaves for bunching sales as the plant grows.

Chard stalks may be green, yellow, red, or multicolored in the case of 'Bright Lights'.

Cool the plant quickly after harvest and keep damp or in high humidity.

CORN

Corn is perhaps my favorite home garden and market crop, and I don't believe

Tomatoes, left, and peppers, right, both will be staked to keep their fruit off the ground and away rain-splashed dirt and soil-borne diseases.

summer really begins until the first silks turn brown.

I know many growers who don't plant corn because of the space it takes, its reputation as a heavy feeder, and what many see as a low return per acre, and there is *some* truth in all that. But I also believe our customers would burn us out if we didn't grow sweet corn for them. We charge 50 cents an ear at the market or farm stand, and I can usually get two ears per plant, so a return of nearly $1 per square foot is just fine. (Others say they get only one ear per plant, but that isn't my experience at all.)

As to corn being a heavy feeder, I've heard so many different estimates of how many tons of nitrogen, potassium, and phosphorous (NPK) an acre of corn requires per year that I finally gave up and just kept growing it. When you combine crop rotation with the use of compost and leguminous cover crops such as alfalfa, clover, or hairy vetch, the corn does just fine. Whenever I start thinking in terms of NPK I get dysfunctional, so I just go out and feed the compost piles instead.

We space our corn on 12-inch centers, six rows to a bed, with beds about 24 inches apart. We try to grow no more than four beds side by side in a long row, and thus waste almost no space. The close spacing also virtually eliminates weed competition, but it does require top-notch soil.

Some have had good success growing a leguminous cover crop such as hairy vetch

Tip

If you've never tried this, try it this year. Make a sandwich out of sliced tomatoes, cucumber, mayonnaise, and a bit of salt and pepper. You'll be happy.

If you've never tried pickling, then plant some pickling cucumber and try pickling on your own. Some great recipes are out there, and it's quite easy. You can also pickle slicing cukes, but pickling varieties can be more fun.

We sow cucumbers thickly in a wide bed and have few problems with weeds. They are always a good seller.

Calorie for calorie, vegetables provide the same amount of protein as meat.

between 30-inch corn rows, mowing it one or more times to keep the grass down, or lightly tilling it. The rows could also be 38 inches, or 48, depending entirely on preference and the equipment you own. A neighboring farmer grows 45-inch beds because he has a 42-inch mower.

This living mulch has many advantages, not the least of which is giving you something besides mud or dusty dirt to stand on while harvesting.

If your field isn't irrigated, then be aware that your cover crop could compete with corn or any other field crop for soil moisture. As with so many other things, it's best to start small and see how things work. See how different cover crops work, whether they can be tilled under lightly or mowed, vary your timing, that sort of thing. Experiment and keep records.

Corn should always be direct-seeded, the depth determined by soil moisture. Since organic growers cannot use fungicides to prevent rot, you may need to wait a week or two to plant, until the morning soil temperature reaches at least 60°F, and 65 is better. If I don't have the first corn at the market, I have the best—and the last, because we sow a number of times, not just once. Harvest when the silks are brown and dry. Remove all crop debris from the field in fall and compost it.

The only pest we tend to get is earworms, but not many. Just here and there. Our customers generally don't care, but some reject ears with worm damage. So we take them home and freeze them.

As to varieties, we grow only 'Silver Queen' because I like it, and our customers have come to expect it. We also grow some ornamental corn to offer in the fall, but we don't sell a whole lot. It just looks good.

CUCUMBERS

Here's where I really get into trouble. Any seed catalog or grower I've ever known recommends growing cucumbers with wide spacing: Seed up to 12 inches apart, rows up to 6 feet apart. In rows, hills, or trellises, it doesn't matter. It's wide spacing everywhere you turn.

Except at our place. We used to grow in hills at the farm, mind you, but then I began to experiment with wide rows, because it worked so well with legumes, root crops, and leaf greens.

Now, one major U.S. seed company—which recommends growing seeds 12 inches apart in rows 48 inches apart—says to expect about 30 pounds of cukes per 25 feet of row. That is 30 pounds of cukes for 100 square feet of land.

We, on the other hand, broadcast seeds to be no more than 2 to 4 inches apart on average in a 12-inch wide row, about 2 feet between rows. And we don't get 30 pounds of cukes per 25 row feet; we get 100 on only 75 square feet of land. With no weeds. (We broadcast by hand and then cover with soil manually.)

I don't know if cucumber seeds like to snuggle up the way peas and beans do, or if my soil is better than most, or both. It can get a little messy what with stepping in amongst the vines in closely-spaced rows, but the system works—for me. Others are wise to stay with what works for them.

I don't have room in the greenhouse for cucumbers, so we always direct-seed. If you do grow from transplants, don't let them get rootbound and remember that all members of the squash family don't like their roots disturbed. Direct-seeded cukes germinate quickly and take over while your back is turned. Just don't rush them; wait until the soil is a good 70°F, maybe 10 days after last frost.

Cukes are a good source of vitamin C and fiber. The skin also contains beneficial minerals, although some find the skins bitter. Pick the fruit every day or two as overgrown cucumbers put a big strain on the plant.

We seldom have a problem with pests, but cucumber beetles are common in many areas.

EGGPLANT

As a child I'd never eat eggplant because I didn't like the sound of it. I didn't like the sound of cottage cheese, either, or yogurt, or pimientos, or olives.

Then Janet made me eggplant Parmesan. Now I just don't eat cottage cheese.

Eggplants are a mainstay in our market gardens because they're so easy to grow and so attractive in a market box or basket. We grow only the purple varieties—both the traditional plump, and the elongated types—because I've found that consumers have decided what an eggplant should look like and that's that. Most customers don't expect white, green, or striped fruit so we don't grow them except to experiment with various varieties for taste.

Eggplant is a hot-weather plant, preferring daytime temperatures of 80°F or more to do really well. Cool weather sets them back for a good while. If your season is too short for them, try planting in raised beds, which warm up faster in the spring.

Grow only from transplants sown eight weeks before setting out. They need 75 to 80°F temperatures to germinate predictably, and in a cool climate should not be set out until the weather has settled, about the time you'd put out peppers. Some prefer to stake their plants to keep the fruit off the ground, which can improve the shape of elongated fruit. If you do stake, it's most economical and speedy to use the Florida weave system.

Tip

The skin of an older eggplant can be tough, so skin those. Fruits will last up to 10 days in a cooler with high humidity.

We've had luck drying eggplant slices for later use.

Eggplant, curled up and fast asleep.

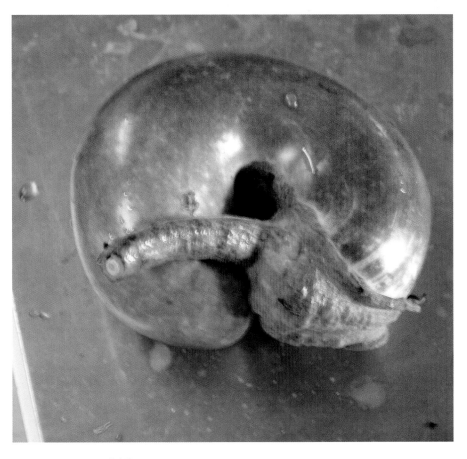

*Like other root vegetables,
garlic loves potassium,
and organically rich soil
normally has plenty.
The plants respond well
to a regular foliar feed
with fish emulsion.*

Eggplant is a good source of fiber and carbohydrates, along with a bit of protein, vitamin C, and iron. Mostly this tomato relative absorbs the flavor of whatever it's cooked with, and has a nice chewy texture.

The only insect pests we have trouble with are flea beetles, especially early in the season. We treat these with insecticidal soap. Pyrethrin or rotenone are also effective. Bt is effective against Colorado potato beetles if you get them.

This long-season crop will have weed problems, and isn't suited for wide beds since it won't smother weeds or set much fruit if closely spaced, so set plants 18 inches or more apart in straight, well-mulched rows 30 inches or more apart. If drip irrigation is used, then set out transplants in double rows (as in a wide bed) through black plastic mulch with drip tape beneath. Stagger the transplants, as with broccoli, to form parallelograms, not squares when seen from above. The black plastic not only heats up the row but

suppresses weeds. Broadcast a cover crop such as rye grass between the rows to keep soil-borne diseases and mud from splashing on to the eggplant. Mow the grass regularly, and turn it under at season's end.

Harvest when the fruit is full-sized but still glossy by snipping off the stem with hand clippers. Dull fruits are over-ripe.

GARLIC

One of the value-added items we sell at the farmers' market is homemade pesto, and we quickly discovered that the taste difference between organic and nonorganic garlic is at least as pronounced as it is for celery. The taste is so good, in fact, that we're able to market an 8-ounce jar of "Janet's Pantry" pesto for $7.95. Those who have tasted it don't blink an eye at the price. The recipe is included in this chapter.

So many types of garlic exist today that it's almost impossible to sort them out. We prefer softneck varieties for their longer storage life to get us through winter, and we also select for

Eggplants come in many sizes and colors, but the standard dark purple varieties sell best for us. Food buyers tend to be conservative.

Nasturtium leaves and flowers have a delicious peppery taste. We add them to our mesclun mixes.

Gourds are fun, they keep well, and you can end up with some really wild shapes and colors.

types that seem to produce the largest cloves per bulb. Small cloves are a dickens to peel.

We always plant in the fall, after a frost but well before ground freeze to get the roots started. These plants will green up in spring and are ready for a good summer crop when tomatoes ripen and customer interest is high. For added value, we braid the tops to be hung decoratively. As with any root vegetable, I make sure the garlic beds are covered all winter in straw or leaves to eliminate any weeding in spring.

Garlic, even more than some crops, thrives on rich, friable soil. It, and other members of the onion family, are said to be highly beneficial to crops following it in the rotation—particularly brassicas—and our experience is that it certainly

doesn't hurt. Like other root vegetables, garlic loves potassium, and organically rich soil normally has plenty. The plants respond well to a regular foliar feed with fish emulsion.

In planting, set the cloves root-end down about 2 inches deep, 2 to 4 inches apart. We grow garlic in 3-foot wide beds and mulch deeply with straw or leaves after planting to keep the beds weed-free in spring. This slows spring growth somewhat by keeping the soil cool, but eliminates the need for much irrigation. Besides, nobody comes to the market in a hurry to get the season's first garlic.

Harvest when the tops begin to yellow and fall over. Don't wait until all greenery is gone or the bulbs may have lost their papery

Kale is a nutritious and attractive cold-weather crop.

Leafy greens in general are rich in vitamins A (from beta-carotene) and C, and are also good sources of calcium, iron, and magnesium.

Tip

Small bulbs can be stir-fried, with larger ones sliced into a salad.

covering and therefore won't keep well. We clean (brush off soil, don't wash), sort, and braid the garlic immediately after harvest. Bulbs with missing or broken tops are stored in mesh bags after cutting off all but an inch of the tops.

Softneck garlic will often store right into next year's harvest, although we seldom have much left by then.

The three garlic types are (1) softneck, which has the most pungent flavor and stores the best; (2) stiffneck, which cannot be braided and has a milder flavor; and (3) elephant garlic, which produces bulbs weighing up to 1/2 pound with a mild flavor. Again, we grow only softneck because of its ability to be braided and stored.

Garlic is widely reported to reduce cholesterol and lower blood pressure, while stimulating the immune system, but studies vary on this. We eat it because we love the taste.

GOURDS

Gourds are fun, and that's about all there is to it. They're grown like winter squash, and look good on the harvest table. Unfortunately, our farmers' market closes Oct. 31, and gourd interest doesn't seem to rise until closer to Thanksgiving. We've sold a bunch of them at the farm stand, but not at market. I don't know why.

Still, they are fun, they keep well, and you can end up with some really wild

shapes and colors, so find an underused piece of land, throw a bunch of gourd seeds at it, and amble over in early October to see what you have.

If we have any that might look good with the fall produce, then we'll put them out in a half-bushel basket. If not, well . . .

GREENS

By greens, I mean the greens that are added to a salad or mesclun mix—not the lettuces themselves. There are countless varieties, from dandelion greens (thought once to be a weed) to purslane (thought also to be a weed) to arugula (a novelty 30 years ago, and now considered a staple in many gardens and farms).

Except perhaps for arugula, pak choi, and tatsoi, these greens are generally not worth growing for sale—only as an ingredient in a mesclun mix. There they are welcome additions, not only for their fragrance and flavor, but for looks as well. Mizuna and minutina, for example, perk up a mesclun mix visually like few other greens.

Leafy greens in general are rich in vitamins A (from beta-carotene) and C, and are also good sources of calcium, iron, and magnesium.

We direct-seed most of ours in a raised bed, sown thickly by hand and covered. We generally harvest with scissors by giving part of the bed a haircut whenever needed. Replant in sections as a crop is used up, bolts, or turns bitter. We plant greens and leaf lettuce alike; we're usually able to get at least three harvests from the same bed, depending on weather. Be sure to irrigate regularly.

We've experimented with a wide variety of greens over the years, both for ease of growing and their worth in a salad mix, and while there are many we haven't tried, we can recommend the following:

Arugula (a must), cress (fast growing), sorrel (intense flavor), minutina (attractive), various mustard greens (for taste), mibuna (for salads and cooking), mizuna (attractive), tatsoi (taste and appearance), New Zealand spinach (taste and appearance), and dill (we add dill to our mesclun mix for the fragrance).

Mâche is one of the best such salad ingredients, so it gets its own section later this alphabetical rotation.

Again, there are many others we like, and many we haven't tried, so don't take this list as marching orders. It's just a solid

Arugula is a hardy annual that makes a lively addition to salads, soups and sautéed vegetable dishes.

Salad burnet is a perennial whose young leaves impart a tangy, cucumber taste to salads.

Lettuce is best sold in clear plastic bags such as these, priced per bag, not pound. This works for anything from head lettuce to mesclun mix. Make sure the bags are spritzed occasionally and are kept out of the sun.

Mint is a perennial herb that can be invasive, and is best grown in a container. It has numerous culinary uses.

Perennial lavender is an excellent cut or dried flower, attracts bees, deters moths, and is nice in potpourris.

Sage, a perennial, is nice with turkey, apple dishes, pork, poultry, soups and stews.

Oregano is a low-growing perennial renowned for its use in Italian cooking.

Thyme is a perennial that comes in many forms, with a lemony fragrance useful in an endless array of dishes, especially meat, fish, and poultry.

Rosemary is a highly fragrant culinary herb, to be used sparingly with lamb and roasts, fish, and rice. Lay a sprig in the linen drawer for fragrance.

Winter savory is a perennial herb used with meat, poultry, and fish.

Hoarhound has long been used in candy, or in tea for help with sore throats or a cough.

Sorrel is a hardy perennial whose succulent young leaves are used in salads or with fish, sauces, and creamy soups.

Kale is an excellent choice to plant after a summer garlic harvest because the leaves improve markedly in flavor after a few frosts, and can be harvested through the winter in many areas.

All parts of a leek are edible, so don't store them too long lest the tops degrade.

selection to begin with. Tastes vary. I didn't name 'Osaka Purple' as a mustard green, for instance, because Janet considers it a weed and she's probably going to read this book.

With some planning, you should be able to find enough varieties to take you right through the season, even in hot weather.

About the only pest for most of these greens is the flea beetle.

HERBS

We've always grown herbs, culinary and otherwise, and while their value as a cash crop is somewhat modest, we've always felt they add another dimension to both the farm stand and farmers' market. Bunched basil, arugula, and parsley (both curly and flat-leafed) are the biggest sellers, but we also tried to put out as much as we had time to pick. That which didn't sell we hang from the farm stand rafters to dry. It is a perfect spot.

The herbs we found to be the most popular were dill, thyme, mint, oregano, cilantro, 'Mexican Mint Marigold' (as a perfect substitute for French tarragon), lavender, summer savory, lemon balm, chives, and marjoram.

All herbs are easy to grow, and in the case of lemon balm, a little too easy. Be sure you contain it.

KALE

Kale, with its gorgeous leaves and husky, succulent flavor, is a nutritional powerhouse, providing more nutrients per calorie than almost any other vegetable. It's an excellent source of vitamins K, A, and C, and fiber, along with a good supply of other nutrients.

It's one of the oldest cultivated brassicas, a significant crop in Roman times and through the Middle Ages. It is a relative of collards, which are grown the same way but have a different leaf. Kale looks so handsome in a field that we try to grow some as close to the farm stand as possible, if only for the visual appeal.

Kale is easy to grow, and is extremely cold-hardy. It's an excellent choice to plant after a summer garlic harvest because the leaves improve markedly in flavor after a few frosts, and can be harvested through the winter in many areas. In colder zones it will last through many a frost. It's possible to plant a spring crop, and we have, but the flavor of the fall crop is so superior that we don't push the spring crop. In hot weather kale turns bitter.

Direct-seed or sow for transplants as you would for broccoli, plants spaced to 8 inches. Plant extra seed to allow for nongermination in summer, then thin if necessary.

The plants grow 2 to 3 feet tall, and leaves can be harvested and bunched for sale, leaving the plant to regrow new leaves.

Be sure not to damage the growing inner leaves when picking. Cool quickly and store as for other brassicas.

Pests and disease are virtually unknown on kale. What more could you want in a crop?

KOHLRABI

Remember how I didn't eat cottage cheese because the way its name sounded? Kohlrabi is the same way, although I'm also suspect about the way it looks. So (ahem) we don't grow it. We tried once, but it just didn't make it as a market crop.

If you want to give it a try, maybe as a novelty, it's another brassica and thus a nutrient powerhouse. Grow them tightly spaced in a bed as you would radishes. It's fast-growing, loves cool weather, and should be harvested at no more than 2 inches or so, then bunched for sale as you would radishes. Kohlrabi has a flavor similar to white turnips, and comes in both white and purple varieties.

The bulb grows above ground, so you'll know when it's the right size. Older ones get woody and off-flavored.

Kohlrabi is susceptible to the same pests and diseases as cabbage.

LEEKS

My earliest experience with well-grown leeks was when I visited the Massachusetts home of James Underwood Crockett (original host of *Crockett's Victory Garden*) for a magazine article I was writing in 1976. The wonder was not so much the size of his garden (it was small; maybe 30x60 feet); the wonder was that it had absolutely no weeds, and had about the most handsome leeks I had ever seen. I recall them being the size of baseball bats ('Broad London' was the variety, now called 'American Flag'), and I have been striving ever since to grow the same leeks.

Success evades me.

Still, I love growing them because it gives me a member of the onion family to enjoy fresh all winter, when their taste is particularly good—a little sweeter than most onions.

The thing to go for in leeks is a broad, tall, edible stem. And that means (here we go again) rich, friable soil.

Crockett direct-seeded his leeks, but transplants yield larger leeks by fall because of their head start. Sow them 8 to 10 weeks before last frost in small-plug flats, or several seeds in a larger cell (our own pref-

Black-seeded Simpson, my favorite lettuce, grows in a wide row.

erence, because they separate so easily and we plant by hand). By the time of last frost and ground thaw, they'll be at least 8 inches tall and about pencil-thick.

After hardening off, we transplant with a bulb-planting tool, a good 6 inches deep (or more) and 6 inches apart in rows 18 to 24 inches apart. This is one root vegetable we don't plant in wide beds, but in rows, because leeks must be hilled up like potatoes several times during the season.

Leave only 2 inches or so of greenery sticking out the hole, and fill in lightly with topsoil. We then mulch heavily with rotting

Tip

Leeks store better in the ground than they do in a cooler. Or, store them in a box with moist soil in a cool place.

Red lettuce is superb in a mesclun mix.
Photo courtesy of Fresh Harvest Co-op

straw or leaves. Once the tops begin to grow, hill them up several times to keep the stalk blanched and long. This can be done by hand or with a potato hiller. Alternatively, the hilling can be done with mulch, the way we hill up potatoes. Use wet, partially decomposed straw or leaves for best results. Dry material can blow away or will settle to become far less than you need.

Don't let weeds get ahead of your hilling or life becomes a mess.

As with other long-season root vegetables, mulch the beds heavily against frost and harvest right through until spring. Harvest at just about any stage, but I prefer a diameter of at least an inch for market purposes. Unlike onions, garlic, and shallots, leek tops do not die back as the crop matures. The top growth, or flag, remains a dark green into maturity.

Leeks can be harvested by a gentle twisting and pulling, or by digging. Much depends on your soil. Don't leave them too long in cool storage or the tops will wilt and rot, which means you can market only the blanched stem. The green tops are edible, but often quite long, so I trim them back for a more polite appearance. It looks good. Leeks also look good with their green flags trimmed in a market basket.

LETTUCE

Lettuce is an organic grower's best friend, not only for ease of planting and harvest, but because of its dollar return per square foot of land, and the fact that you can grow it pretty much all winter in an unheated greenhouse.

And, I can tell you this: There are few things a good restaurant wants more than fresh, organic salad greens any day of the year.

Iceberg has long been America's choice in lettuce but we don't grow it. For one thing, outside of a little fiber and vitamin A, it has virtually no food value. It also looks like a green bowling ball, ties up soil for longer than we'd like, and is extremely sensitive to heat. An early hot spell can ruin a crop.

Supermarkets are full of iceberg lettuce from California, so we grow what the supermarkets *don't* offer—and that list is almost endless. It includes red and green romaines, butterhead, Bibb, crisp leaf lettuce, and a wide variety of soft leaf lettuces.

We sell some lettuce as heads, but prefer to offer a mesclun mix (meaning a salad mix with a variety of greens and perhaps herbs). This has been equally popular at the farm stand, the farmers' market, restaurants, and in specialty stores (such as a local fish market and general store). We select greens for the mix based on flavor, appearance, and what's in season. These are far superior to supermarket mixes, and enable us to price at $9 per pound in a clear bag. (Sell by the bag, not the pound.)

Hot summer weather is a killer for lettuce, so if you can't supply restaurants through midsummer heat and dryness, tell them so there won't be any hard feelings. It's better to take a break than to offer anything less than the best quality. Take it up again in September.

Most lettuce is a fast-growing cool-season crop that can be planted any time the soil isn't frozen. Seed will germinate at temperatures as low as 40°F. It grows best at temperatures below 70°F. A crop can generally be picked for about three weeks. After that quality suffers, even if the plant isn't bolting. Always taste-test anything you pick.

Romaine is both nutritious and heat tolerant. Butterhead (Boston, Bibb) varieties are also nutritious, but like cool weather. Loose-leaf lettuces are also nutritious, are fast-growing, and are more heat-tolerant than heading types.

We grow head lettuce as transplants, sowing seeds three to four weeks before hardening off and setting out. Once prepared for the change, seedlings can tolerate a hard frost easily. We keep one end of the greenhouse in shade cloth to protect the seedlings on a hot, sunny day. Stagger plantings for a steady supply through spring and early summer.

Plant heat-resistant cultivars as the season warms up, with plenty of irrigation for their shallow root systems. Lettuce seed germinates poorly or not at all in hot weather, above 75° or 80°F, so germinate your seeds in flats in a cool cellar, or get them to begin dormancy by rolling a batch up in wet paper towels in the fridge for four or five days—don't use newspaper; it falls apart in a mush. Then transplant the sprouted seeds to flats to be transplanted a few weeks later.

I grow leaf lettuce in wide beds, direct-seeded, watered, and covered with soil. The lettuce doesn't seem to mind being so thickly planted that it looks like a green carpet. The system not only suppresses weeds and keeps the soil moist, but keeps the lettuce cleaner without as much dirt splashing up from rain. Again, that violates most rules but it's worked for me for years.

Always harvest lettuce in the early morning before the sun gets on it. The plants are 90 percent water, and can lose moisture very fast. We harvest leaf lettuce with long scissors.

I like to harvest head lettuce at a small stage, about a 4-inch diameter, when it looks more appetizing than a big head. People who won't buy a large head will buy

two small ones weighing about the same. I don't know why.

For winter growing, about all you need is an unheated greenhouse and some secondary protection from hard and prolonged freezes, such as a floating row cover or low poly tunnels.

MÂCHE

Not everyone knows of this vegetable, also known as corn salad or lamb's lettuce, and that's too bad because it's tasty and extremely cold-tolerant. It's not actually a lettuce, but its firm leaves make a fine salad by themselves or with other greens and vegetables.

Mâche is a poor germinator so plant extra in anticipation of thinning, and don't

Melons aren't the easiest, most space-efficient crop to grow, but customers love them, and the taste beats anything you'll find in a store.

Okra, after being graded, is weighed out in plastic bags for inclusion in CSA boxes.

Mulch closely with wheat straw, leaves, or weedless hay to keep down weeds and keep the melons clean and free of soil-borne diseases.

expect to see the seedlings for nearly two weeks. It can be planted in the fall and over-wintered like spinach, or sown in early spring for a later crop.

Mâche was a field weed centuries ago, and still grows wild in Europe. I've not seen it in the wild here. Most of its production in France comes from the Nantes region (of Nantes carrot fame) and its success reportedly stems from the fertile, sandy soils of the Loire River.

Mâche should be harvested at less than 3 inches. Larger plants lose their flavor.

MELONS

Melons are summer's reward for hot, humid weather and I love them. Many new varieties require a shorter season, but they still need at least three months of warm weather. I've grown muskmelons and Crenshaws (what a taste!) as a market crop, but have found some of the gourmet French varieties to be overrated.

Melons require a long season of warm weather, and cannot be exposed to a chill even as seedlings. If they are, they won't set fruit. It's best to grow from transplants (grown at 80 degrees F or warmer) that won't be put out until the soil has warmed to at least 75 to 80 degrees F is better. You just can't rush these babies. Sow in cells four weeks before setting out, and don't let the seedlings get rootbound. This means you'll be sowing melons inside when other crops are being direct-seeded, but that's okay.

I plant in hills so the group is easier to mulch with straw, but rows are good when black plastic is used. Set three to four plants in hills 6 feet apart in a row, or 2 to 3 feet apart in rows 6 feet apart. Watermelons will need much more space than muskmelons. Treat seedlings extremely carefully when transplanting, and give them a good helping of worm tea, compost tea, manure tea, or fish emulsion.

We mulch closely with wheat straw (because we have it), leaves, or weedless hay to keep down weeds and keep the melons clean and free of soil-borne diseases. It also helps keep soil moisture levels even, which the melons like. The vines may look sturdy, but they're actually quite sensitive to being moved or stepped on.

Male blossoms will arrive first, followed by female flowers. Each vine will produce about three to four melons.

Powdery mildew is a common problem in wet, hot weather so cut off any affected leaves lest the mildew affect a melon's taste. Bacterial wilt also affects melons; control this by controlling cucumber beetles and aphids.

Harvesting melons is an art form because so many varieties need harvesting at different stages. Muskmelons are harvested at full slip, when the stem almost falls off the fruit. For other types, experience is often the best guide. And read catalog information.

Melons can be stored at near freezing with moderate humidity for two weeks or longer.

Onions are divided into short-day and long-day varieties, depending on how far north or south you grow them.

OKRA

When my brother in coastal Maine last year proudly proclaimed to me that he had successfully grown okra, I said, "But Les, why would you want to *do* such a thing?"

Janet loves okra. Me, I hold it right up there with cottage cheese. It's a Southern treat, however, and has been popular for years in gumbo, soups, stews, deep fried, stir fried, and pickled. It can also be eaten raw.

Okra is a wonderfully attractive plant—tall, with beautifully soft, yellow, hibiscus-like blooms all summer. The fruit comes in green and red, but it all cooks (or pickles) down to green. One serving provides a good amount of vitamins A and C, and some calcium.

We grow from transplants, sown three to four weeks before setting out after danger of frost. Some recommend a 12- to 18-inch spacing, with 36 inches between rows, but I prefer to give each plant at least two, maybe three feet of space in all directions to encourage side branching.

In areas with long summers, okra can be cut back to 8 to 10 inches in late June after the spring crop is harvested to produce an often even larger fall crop.

Harvest okra when it's small; 2 1/2 to 3 1/2 inches. Larger fruits get tough. And use some caution when picking okra because most varieties have tiny hairs on the plant that can leave your skin itching for a long time. Use gloves and long sleeves to remove the pods with hand pruners. Okra stores well at cool temperatures and high humidity for a week to 10 days.

Pests and diseases are seldom a problem with okra. It is also drought-tolerant and will need no irrigation. It just chugs along all summer producing these little pointy things that I won't eat.

ONIONS

I fell in love with onions for the first time when we planted and sold 'Walla Walla Sweets,' an onion you can eat like an apple. We started with 2,000 and doubled the planting until we finally reached a glut limit of about 6,000 onions at the farm stand. Walla Wallas won't keep long, so the few that we didn't sell we chopped up and froze.

These onions are not available in stores, so once word got around, customers would come to our farm for the Walla Wallas alone, which we sold at a premium—more than 'Vidalias,' because they were local and fresh.

On moving to Tennessee we ran headfirst into the long-day, short-day onion phenomenon, which is the first thing every grower must know before planting. In a nutshell, long-day onions only grow in the North, and short-day onions only in the South, with a dividing line at about 36° latitude, or roughly the Kansas/Oklahoma border.

The reason is this: Onion bulbs don't begin to grow until their tops are fully mature. The bigger the top, the bigger the bulb. The top stops growing when the bulb begins forming, and that change is triggered by day length. A long-day onion, such as the

Walla-Walla sweet onions, a long-season variety.

> ### Tip
>
> Even if I don't sell many onions, I love to grow them because they look so good on the harvest table. Try to grow varieties that customers aren't apt to find in stores, or in sizes they won't find there.
>
> The pungency of onions repels many insects, and most crops, especially brassicas, seem to do extremely well following onions in the rotation.
>
> The tears you shed when slicing an onion are tears of joy because this vegetable is a good source of potassium, vitamin C, fiber, and is rich in cancer-fighting antioxidants.

Not always thing of beauty, but parsnips make a lovely sweet, tangy dish. Growth of this one was interrupted by heavy soil.

Parsnips, like carrots, are excellent root cellar keepers—stored for months in an above-freezing place where they can't dry out.

Tip

Definitely include a recipe with your parsnips. This will increase sales until a familiarity develops for the vegetable. Many people we've found will snap them up as a memory food—their mothers or grand-mothers cooked them—but they won't know quite how.

Walla Walla, forms a bulb only after June 21 (the longest day) wanes into shorter ones. Boston, on June 21, has a full hour more daylight than Atlanta, so the tops have plenty of time to grow and feed a big bulb. In the South, a long-day onion will form a bulb when the top is far too small, so you get a big scallion.

Short-day onions, such as the Vidalia, begin bulb formation with far less daylight, so they can be direct-seeded in the fall. By spring, when bulbing begins, these onions have produced large-enough tops to support large bulbs, which mature through midsummer. This is important to know, because many books and seed catalogs don't mention it.

Onions prefer sandy, loamy soils rich in humus—well drained and with plenty of sun. Heavier soils should be amended with compost well before planting. Onion seeds are slow to germinate and grow, and bulbs won't tolerate weed competition, so the bed should be essentially weed-free and remain that way.

As with other root vegetables, we prepare our onion beds the fall before, by mulching them deeply in straw or leaves, then pulling this back for spring planting. This virtually eliminates spring weeding.

The important thing is to encourage your plants to grow large tops as quickly as possible. Once bulbing begins, the bulb's size is already determined by the top's size,

so there's no need for fertilizer or soil amendments after that.

Onions grown from seed are cheaper, and you can choose from far more cultivars than if growing from sets or plants. In northern areas, onions may be direct-seeded in early spring as soon as soil can be worked. Or, start them in a greenhouse four to six weeks before setting out. In milder climates, onions may be direct-seeded in the fall.

When direct-seeding, sow thickly in a band several inches wide, seeds 1/2 inch deep and 1/2 inch apart, with bands 12 to 24 inches apart, depending on your mulching or tilling method.

Interplant with radishes if you want to mark the bands. Thin seedlings anywhere from 2 to 4 inches apart, depending on what you're growing. Bunching onions can be closely spaced; larger ones need more room. We like to grow both, along with green onions (or scallions) for early spring sale. Johnny's Selected Seeds, and perhaps other companies, now offer a deep purple scallion, so we'll probably try at least some of these to distinguish our offerings from others.

Supply your onion beds with an inch of water a week, and it's far better to spread this out so the roots always have enough moisture to feed the tops and bulb growth.

Harvest bulbs when the tops are dry and falling over. In warm, dry weather, dig the onions and let them cure for at least several days in the field, up to a week. This is essential for long storage. Some say cut the tops off first, but I've found it doesn't matter.

Store onions in mesh sacks or where they can receive good air circulation in the shade.

PARSNIPS

Parsnips have a lineage dating back to Roman times; they were a staple long before potatoes, and were used as a sweetener when honey was scarce. Today they can be hard to find. This is a shame because the sweet, nutty taste of fresh winter parsnips is a fond memory in our house. Moreover, they're a good source of carbohydrates, vitamin C, minerals such as potassium and calcium, and fiber. Better than potatoes.

We grow them for ourselves, but also like to grow them for sale in spring when the vegetable offerings are somewhat lean. It may take a while to get a following, but my feeling is that consumers will want to buy them more than you'll want to grow them.

Parsnips are grown much like carrots, and in fact look like an albino carrot; they can be eaten fresh or cooked. But they need a long growing season (up to 130 days), deeply tilled soil, and a long germination time. We think it's worth it.

Many books, seed catalogs, and websites state flatly that parsnip soil must be tilled to a depth of 2 feet, but we've found that unnecessary. Nor does the soil need to be nitrogen-rich; that causes root forking. We sow in clay/loam soil loosened to less than 12 inches and harvest 6- to 8-inch parsnips. That's all we want.

Sow in spring to early summer in two rows 12 inches apart in an 18-inch bed, seeding thickly 1/2 inch deep, then thinning to 3 inches apart. Seed can take three weeks or more to germinate and need constant moisture during that time, so we cover the beds with grass clippings, leaf mold, or compost. Don't allow the soil to crust or seedlings may not get through it.

The deep roots are reasonably immune to dry spells, so weed control is all that's required until winter. Parsnips sweeten wonderfully after a few frosts, so you can harvest all winter (under mulch in cold areas) and right into spring. Harvest before the roots send up new tops, lest they lose flavor.

Parsnips, like carrots, are excellent root cellar keepers—stored for months in an above-freezing place where they can't dry

out. I know of a Dutch grower who harvests carrots when they're moist, with soil on them, covered with another layer of soil when they're in the crates, keeping them up to a year this way. No reason why this wouldn't work with parsnips.

PEAS

We haven't grown garden peas since 1979 when Burpee introduced the sugar snap pea, and our customers are grateful because nobody has time today to shell peas, and yet they love them. No matter how many sugar snaps we come to the market with, we never leave with a single one. And we can charge a nice premium.

As a grower I am especially grateful for the change, because for the small inconvenience of erecting a trellis I get the wonderful

Supermarkets charge a premium for yellow and orange bell peppers because of the color alone. It's worth trying.

The market for ethnic peppers, such as this habanera, is growing by leaps and bounds.

Super hot peppers, such as these Super Chiles, a 1988 All-America Selections winner, are more in demand today, but don't sell as well as the milder varieties.

convenience of picking standing up with a concentrated harvest that can be spread out over weeks with staggered plantings. You can't stagger sugar snaps the way you can some vegetables, because I've found they all want to come in at about the same time, but we can get a good harvest range.

The only problem with peas is deer. They will wipe you out in one night. If you have deer, you need either a deer-proof fence, an outside dog, or, well, that's about it. Any deerproofing methods I've ever heard of won't work with peas.

As we break the rules for many things, we break the rule for peas. Instead of planting in a row or narrow band, we plant in an 8- to 10–inch wide bed, sowing seed 1 inch deep and 1 inch apart or less. I've found that peas love crowding, and do far better that way. We arrange trellises on the south or east side of beds, so the peas will grow toward it (and the sun) instead of some other direction.

Plant as soon as the soil can be worked in the North. In Tennessee, where my ground doesn't always freeze all winter, I begin planting in late February for a market harvest in May.

We plant on trellises comprised of 4x4s and a 6-foot stock fence because a trellis full of peas is heavy. We also grow along a 4-foot stock fence because it's there.

Today snap peas are available in almost-stringless form, as well as bush types and earlier yielding types and sugar snap, so we're experimenting with these. But we're not giving up the trellises.

As for snow peas—the sweet, flat-podded peas from which sugar snaps were developed—we plant just enough of these to use ourselves and fill a very modest market demand. In college towns and near urban areas, the demand tends to be some-what higher.

Aside from an occasional deer, we've never had problems with pests or diseases on our peas, except for an occasional visit from fusarium wilt, which crop rotation pretty much eliminates. We like to follow a pea crop with corn, one of the brassicas, or greens.

PEPPERS

Peppers are America's second-favorite vegetable next to tomatoes, by all accounts, and they should be a grower's favorite crop as well. This warm-weather vegetable is relatively easy to grow and provides a sustained harvest right up until frost. Peppers take forever to set fruit, it seems, but once they do it's hard to keep up with the harvest.

Bell peppers are the runaway favorite, and these sell well in the green stage, but sweetness and nutritional value improve greatly if peppers are left to mature into red. Colored peppers should also be left to mature into their colors of red, orange, yellow, purple, even white these days. Just be sure not to let any overripe fruit remain on the plant lest productivity take a nosedive. As with so many flowers and vegetables, once a plant knows it's set fruit and seed, it's time to fold up and relax.

Bells are only the tip of the pepper: a wise market grower will also provide plenty of banana peppers, chiles, jalapeños, and other specialty peppers. We avoid growing the super-hots, such as habaneras, because of problems associated with handling them (burned fingers), but that's just a personal decision. In many regions of the country, ethnic super-hots sell extremely well, so consider your market in deciding what to grow.

If you run a CSA, be sure to let customers know what they're getting if you supply a box with hot peppers. A simple label printed out on a sheet is all you need. It might say, "The wrinkled, red, lemon-sized thing is hot and should be approached with caution."

We grow peppers, like tomatoes, only as transplants. We sow directly into 2 1/4-inch cells because (1) I hate transplanting, and (2) I'm convinced that a large cell provides better root growth and stronger plants. Sow a good eight weeks before setting out. We germinate peppers in the warmest part of the greenhouse, near the heater, then move them (or redirect the heater louvers) at the seedling stage.

Our Johnny's seed catalog suggests that plant production increases if young plants with three true leaves are subjected to four weeks of cool temperatures—53 to 55°F—

Summer bounty, just in from the field.

in which case flats should be sown a week or two earlier than normal. I've not tried this as we don't have a place to do it, but thought it worth passing along.

Transplant after night temperatures remain above 50°F. We set pepper transplants 18 inches apart in two-foot rows, staggered just as with brassicas. Black plastic and drip tape is especially worthwhile with peppers because of their need for warmth, steady moisture, and a long growing season. Alternatively, we prepare the beds well, mulch heavily with straw, and sow the transplants into that.

Remember to grow not only green/red bell peppers, but red, orange, and yellow bells as well. The supermarket charges a premium for these, and so might you.

> ### Tip
>
> Bell peppers may be diced into fingernail-sized pieces and frozen in airtight plastic bags. They lose crispness, but not flavor. Thaw them to use in any way you would use a fresh pepper.

Washing potatoes. We prefer varieties with shallow eyes for ease of cleaning.

We discovered years ago that the answer to Colorado potato beetles is to hill up potatoes not with soil but with last year's straw or leaves.

We don't grow purple peppers because I don't want customers being confused with our eggplants. Most decisions a buyer makes are based on a two-second look, and confusion doesn't pay. A white-ripening-to-scarlet pepper is available, however, and might be worth a try.

Peppers don't need much nitrogen lest the plant produce more leaves than fruit. They do need abundant calcium and phosphorous, however. A calcium shortage may cause stunted growth, bleached or curled leaf tips, or blossom-end rot, but the "shortage" is more likely an unavailability caused by acid soil.

We rarely have pest or disease problems with peppers. Cutworms are eliminated by mulching, and we've found that spreading fresh grass clippings around new transplants either repels cutworms or confuses them. Either way, it eliminates the problem.

Harvest bell peppers when they're firm and 3 to 4 inches long. They can be pulled from the plant or cut. Pulling is quicker, but be careful not to damage the plant. Cool temperatures and high humidity will keep peppers fresh for several weeks. The

moment a pepper loses its gloss and starts to soften, compost it.

Green chiles should be harvested at the green stage since many recipes call for that. Jalapeños also are generally harvested green, in part because it takes another three weeks for them to turn red. And most customers identify them by the green color.

POTATOES

The organic grower may face uncertainty on the question of whether to grow potatoes. We've had great success growing them pest-free, and particularly love to sell a crop of early "new" potatoes. We get a premium for these, and customers know they're worth every penny.

The question arises as to whether to raise a main crop or to turn the beds over to some other use. If you have the equipment to plant, hill up, and harvest mechanically then a main crop is worth the effort, particularly since there are so many varieties from which to choose. On a smaller scale, however, much depends on your market and the competition. Potatoes are a great addition to a CSA basket because

Tip

Include with your potatoes the simplest, most delicious recipe you can find to help spur purchases.

customers expect a nice diverse offering. But main-season potatoes are not fast sellers at our farmers' market.

To further confuse matters, we discovered years ago that the answer to Colorado potato beetles is to hill up potatoes not with soil but with last year's straw or leaves. Since doing that we have never had a problem with beetles, scab, weeds, or anything else. The mulch seems to level out heat and dryness cycles that may stress the plants. But I know of no good mechanical process for spreading leaves and straw on potato beds, so this method is left to the small grower.

For us, the answer has been to keep growing them. We sell enough to make the effort worthwhile, I love the plants and blossoms, and the mulching is double-barreled in that it not only helps the potatoes but any crop that follows. And, a bushel of potatoes or onions gives me the soulful satisfaction of seeing a properly stacked woodpile in autumn.

We select varieties for color, taste, and ease of cleaning. We don't want tubers with deep eyes. Our favorites include:

Russets: Shallow eyes, good texture, for baking and mashing. The most popular potato in America.

'Yellow Finns': Yellow-fleshed gourmet potato with a unique sweet buttery flavor for boiling, salads. Stores better than red potatoes or 'Yukon Gold'.

'Yukon Gold': Extremely popular, sweet-tasting potato, keeps its golden color when cooked; for soups, mashing.

Red potatoes: For boiling and salads, few and shallow eyes, good "new" potato, won't store well.

When "eyes" are cut as seed potatoes, have two on each piece. Then "green-sprout" your sections to get potatoes two weeks ahead of anyone else. To do this, put them in darkness for several weeks until sprouts appear. Then expose them to at least eight hours of light for two weeks until short, thick, sturdy, green sprouts form. They're ready to plant when the sprouts are no more than an inch long. European growers have done this for years to get the best price on early potatoes.

You'll want to experiment with timing and site, so don't put all your potatoes into one green-sprout basket the first year.

Plant green-sprouted potatoes only a week or so before last frost date.

Nonsprouted seeds may be planted two weeks or so earlier. Plant in fertile, friable soil, 3 inches deep and 12 inches apart in rows at least 2 feet apart (for ease of mulching if nothing else). In our clay loam we raise the beds to ensure good drainage. Slightly acidic soil is said to help prevent scab—brown hard spots on a tuber's surface—so don't plant in recently-limed areas.

Let the plants grow to 12 inches, then cover all but the upper 3 to 4 inches with mulch. The leaves or straw should be partially rotted and moist, not dry. Hill again in another few weeks.

Harvest baby potatoes about two months after planting. We used to feel around under the mulch to decide on harvest time, but found that we'd often feel marble-sized potatoes and put off harvest, only to find potatoes the size of softballs growing a few inches away.

Harvest main-season potatoes two weeks after the vines have died back naturally to allow skins to set. If you want to harvest earlier, when the potatoes are smaller, simply pull or cut off the plant tops to stop growth.

Dug potatoes should be moved out of the sun and high temperatures. In Ireland's mild climate, potatoes are stored in "clamps" in the field. These are shallow pits covered with hay, straw, or other vegetation, enough to shade the tubers.

Ideally, tubers should be cured at 45 to 60°F and high humidity for two weeks. Then grade the crop, discard any culls, and store in a cooler (38 to 40°F) with high humidity.

Experiment with timing and site, so don't put all your potatoes into one green-sprout basket the first year.

Pumpkins sold like hotcakes at our roadside stand and in the field, but never as well at the farmers' market.

PUMPKINS/WINTER SQUASH

I've had mixed luck selling pumpkins at the farmers' market, but excellent fortune selling them in the field. Families find something soul-satisfying about walking through a field of ripe pumpkins, and we find it satisfying to not have to lug all those babies up to the farm stand. We just clip the stems and leave them in place.

A grower in Maine went to the trouble of scratching into the skin of his small pumpkins the initials of every youngster in town. He let word of that get around, and was sold out without batting an eye.

For our part, we had trouble selling misshapen pumpkins until we put up a sign marketing them as "Ugly Pumpkins." People seemed to feel sorry for these poor things and paid us to take them home.

I tried growing a monster pumpkin one year but the local football team liberated it and left it in the middle of the football field before the next day's game. I should have carved Plum Hill Farm into its skin.

It gets worse. We kept raising our prices above those at the local supermarket and the other local grower (who didn't grow his own pumpkins, but had them brought in). Prices kept climbing up over the years until we began to fear that some folks in town might not be able to afford a good pumpkin. So we put up a sign announcing "Unpriced Pumpkins: pay what you can afford, or pay nothing if that's what you have."

It was the honor system when we were out in the field, and we made a bundle that year.

Getting back to growing, we get the best response from our old favorites, 'Connecticut Field' and 'Howden', to sell as carving pumpkins. We also try to grow a lot of 'Rouge Vif d'Etampes', the so-called Cinderella pumpkin, because it's so showy with bright orange flesh and the squat, sat-on shape of Cinderella's carriage.

We direct-seed all pumpkins, timing the sowing to coincide with a mature crop by the first week in October, just before first frost. Those who will be shipping pumpkins will want them ready by the end of September. Plant 1 inch deep in moist but not wet soil, in which seed will rot. Rather than thin, which takes more time than I ever have, sow seeds 24 inches apart for large pumpkins, 18 inches for small, and hope for the best. We do this as a grid, not rows, because I till between the plants in both directions.

We till once before planting, and then just enough to keep the soil crumbly until the plants become established. Pumpkin leaves will smother most weeds after this, but my larger plan is that my tilling and their shading keep the soil from crusting. This way roaming vines can root second-arily to outrun vine borers. For this reason also, I don't fertilize the main plant, but instead make sure the entire field is fertile for this secondary rooting. I want well-watered, fast-growing vines in the field.

Cut stems off about 2 inches or less from the fruit, because long stems on a pumpkin are

Tip

Radishes need to breathe, but rather than pack them in berry boxes, to save a penny we sell them in open snack bags. They stay moist and look terrific that way.

Flea beetles can damage young leaves, so watch for them and get them early.

apt to break off and not be marketable. I leave winter squash to dry in the field for four or five days before collecting, and try to avoid any exposure to heavy frost, which can lead to rot.

Vine borers prefer 'Blue Hubbard' squash vines to my 'Connecticut Field' pumpkins, according to a University of Illinois study, so I throw some 'Blue Hubbard' seed in the hopper with my pumpkins to use as a trap crop.

Squash bugs are common in our winter squash fields, but I consider them an end-crop annoyance. Ugly, smelly, creepy, but not really crop-ruinous.

If we get many 'Blue Hubbards', we sell them, but I don't know of any winter squash better tasting than butternut, so we grow mostly those, along with the acorn variety because people expect them.

For years I thought no butternut squash could top 'Waltham Butternut' for taste until last year when we plowed up an area near the house that hadn't been cultivated in at least 20 years. After preparing the bed, up came a rush of maybe 100 squash seedlings in an area the size of two bathtubs. We let them grow, without thinning, and got the biggest, tastiest crop of butternut squash I've ever seen—probably 130 pounds' worth with mostly long necks. Clearly some long-buried seed was turned up in the plowing, so now we'll save the open-pollinated seed and maybe modify our squash-planting program next year.

It's as though Nature was saying, "Hey, you got it right with cucumbers. Now try it with squash." After all, that's how squash reseeds—on its own.

RADISHES

Radishes are spring's tangy farewell to winter, and make a welcome offering in a market basket or table. About the only key is to grow them moist-wet and quickly, in fertile soil. I broadcast them in beds with our other root vegetables to be covered with compost, or drill them in rows 2 to 4 inches apart atop or alongside the carrots and such. They must be thinned if broadcast, because young radishes tolerate no competition from anything.

They can be grown any time of year, but I prefer the taste of radishes grown in cool weather and many varieties won't grow in summer's heat and dryness.

As with beets, the greens are far more nutritious than the root itself, so we use them liberally in our mesclun mix. We bunch radishes for sale the same day, keeping them moist with a sprayer lest the expiring leaves hurt the root quality. Topped radishes will last several weeks if kept near freezing at high humidity.

Flea beetles can damage young leaves, so watch for them and get them early. Or rather, *expect* them and get them early. We use insecticidal soap.

As with so many other vegetables, our best-selling radishes are those that look like a

Bi-colored 'Zephyr' summer squash make an attractive display.

radish: red, or a little white, and round. 'Long French' radishes mature more quickly, and daikons are more adaptable as food, but it's those round radishes that sell best.

RUTABAGAS

Most people have heard of this root brassica, but never met one. I was introduced at an early age since it's a cool-weather crop favored in Scandinavian countries, and my heritage is Norwegian. Much of what Norwegians eat is best left out of cookbooks, mind you, but the rutabaga is not one of them.

The white root is also quite nourishing, high in fiber, vitamin C, and endless minerals. We always try to have some available.

The crop takes 100 days or more to grow from seed, so direct-seed outdoors about 15 weeks before first frost, so the roots will mature in cool fall weather. Sow as for any other root crop, but keep in mind that rutabagas can grow to six pounds or more, so space seeds out to 6 inches. Seeds can take three weeks to germinate, so prepare the bed accordingly or use transplants.

Root maggots and flea beetles can be common pests, but we don't see much of either. The late planting time helps with pests. No diseases are usually encountered.

Harvest after the plants have enjoyed a few good frosts, or leave in the ground, mulched against ground freeze, for all-winter harvesting. You may have to loosen the soil before pulling out by the tops. They will store for a month in a fridge, or for four months at 32°F and high humidity, but don't let them freeze.

SPINACH

This is one my favorite vegetables cooked, in salads, in soups, in sandwiches, and raw. It's always a good seller, and—as with our mesclun mix—we sell by the bag, not the pound.

Its reputation as a health green is well deserved, as it contains the highest level of vitamins A and B2 of any vegetable, with good amounts of iron, calcium, and protein.

It grows extremely well in our rich clay/loam soil, but needs a pH of 6.5 or higher in soil with a high nitrogen content and even moisture. Despite appearances, spinach has a deep tap root, so soil that's loose up to 12 inches down will produce best.

Despite its reputation as a spring vegetable, spinach can be grown successfully much of the year with proper irrigation. Germination can be erratic in hot soils,

so we switch over to New Zealand spinach at that point, which has its own delicious flavor raw or steamed. Or, you can freeze the seed for a few days, then moisten it and put it in the cooler for a few days to aid germination. Toward summer's end, begin planting again until just before first frost. A low poly tunnel or floating row covers will protect your spinach crop well into winter.

Spinach germinates and grows quickly, so plant thickly in beds as soon as the soil can be worked and let it shade out weeds, harvesting with scissors. Sell it in poly bags; we include some in our mesclun mix. It's almost impossible to get grit off a spinach crop for some reason, so we grow

Summer squash must be picked small, before seeds develop.

We transplant tomatoes by laying them on their sides and covering all but the top few inches with soil.

flat-leafed varieties and wash it three times, just as we do for some lettuces. We've tried spreading leaf mold on the beds to reduce splashing, and that helped a bit, but we run out of that wonderful material by summertime.

SUMMER SQUASH

Summer squash is one of my favorite crops for ease of growing and harvest. My favorite types are white or yellow patty pan, yellow straightneck, and a yellow/pale green variety called 'Zephyr'. We also grow yellow crookneck for its wonderful flavor, but it has to be picked small or will harden up.

Culture for summer squash is essentially the same as for winter varieties, except that summer squash is at its most flavorful when immature. Some vendors at our market pick at a bare 2 inches, but we prefer a good 6 inches and will allow more. After that, squash tends to get too seedy for pleasant eating.

Summer squash is virtually immune to the predations of vine borers, but make a fine home for squash beetles. Again, we don't much pay them a mind.

SWISS CHARD (SEE CHARD)

TOMATOES

Tomatoes are not only America's favorite garden vegetable, but your best friend as a market grower because the difference, I believe, between organic, locally-grown tomatoes and the supermarket variety is greater than for any other single vegetable. Our customers every year ask when the tomatoes are coming in, and it's never early enough.

The fruit is rich in vitamins A and C, calcium, fiber, and the important antioxidant lycopene. A tomato's health properties may be improved, not diminished, by cooking, according to several studies.

Tomatoes make an excellent hoop house crop, and we always try to supply the hungry folks with 'Early Girl' or some other variety that's both early and reasonably large. We also grow a yellow Mennonite cherry tomato whose name has been lost in the dusts of time, but cherry tomatoes are so crack-prone and take so long to pick that I don't offer many, and charge a premium for those we stock. They make a mockery of supermarket cherries, incidentally.

We also grow 'San Marzano' salsa/sauce tomatoes because they're so rugged and have such a wonderful taste. But beyond that, we grow only beefsteak varieties and we grow them only for taste, nothing else.

That's what makes farmers' markets and CSAs so remarkable. You can grow a tomato for nothing but taste.

In my case, that means 'Brandywine'. It's one of the world's uglier tomatoes: squat, high shoulders, pink not red, off balance usually, but with an absolute symphony of taste. It's so good that we offer samples to non-believers. Coming in a close second are 'Arkansas Traveler', 'Prudens Purple', and a Southern favorite called 'Bradley'. We grow them all, and try others, but these are the sellers.

Outside of 'Early Girl', we also grow only indeterminate varieties and prefer the heirlooms because we can label them as such to draw interest. Heirlooms consistently produce less fruit than modern beefsteaks, but we think it's worth the price. We've found them no more disease-prone than others.

The trick to successful tomatoes is not to rush them. No matter when you plant, they're not going to take off until hot weather arrives. Plant too early in the greenhouse and you'll end up with stressed and rootbound plants that won't produce well. We sow in 3 1/4-inch pots about seven weeks before setting out. Even a light frost is deadly to tomatoes, so take no chances with early transplants unless you have floating row covers.

If frost does hit a crop, irrigate it immediately with sprinkler water. That saved us enormous heartache once.

We transplant tomatoes by laying them on their sides and covering all but the top few inches with soil. They'll root off their stems and get off to a faster start this way. Don't worry that the leaves point sideways; they recover in a day.

We also trellis all our indeterminate varieties in one of two ways. One is on 4-foot stock fence, which we have in abundance. The other is to sink 4x4 posts in the ground with sturdy cross supports between them at a 6-foot height. Space your plants 24 inches apart, and hang a piece of sturdy twine down from the cross support to the base of each plant. Prune tomatoes to a single stem, and wind this around the twine as the tomato grows. The plant's weight keeps the twine in place, and it doesn't cut sap flow.

It is said that tomatoes will fruit earlier if the side shoots, or suckers, that arise between the main stem and leaf axils are removed, but this takes more time than most people have, including myself.

Another interesting practice to keep a plant producing longer is to lower the bottom part gradually to ground level once it has stopped producing, while keeping the top few feet of the plant vertical. When the bottom makes soil contact it takes root and reportedly keeps the top section producing into late fall. This idea is from Eliot Coleman, a man of many good ideas, so it's worth passing along.

We've experimented with allowing our indeterminate varieties to sprawl, rather than being trellised, and it simply doesn't work. They produce more fruit, as the old-timers say, but even when sprawling on a bed of straw, too many tomatoes rot on the ground or are eaten by slugs. And they're hard to pick. Tomato cages likewise are a waste of time, as they're simply too small and fragile to support a well-grown tomato plant.

Tomatoes need abundant soil phosphorous and a neutral pH to prevent calcium from being bound up in the soil. They don't need a nitrogen-rich soil because that will promote foliage over fruit.

Irrigate your tomato crop regularly. A dry spell followed by overwatering can result in fruit literally outgrowing its skin, and splitting. Regular irrigation will also prevent blossom-end rot.

The most common tomato pest, for us, is the tomato hornworm. Once they're spotted, the damage is usually done, and parasitic wasps don't generally get to them in time. The good news is that when you plant abundantly, they're not much but a nuisance. You'll get a few, but I've never

The Brandywine tomato's beauty is in its taste, not its shape.

seen an infestation, and therefore have never dealt with them.

TURNIPS

These root vegetables, another member of the magnificent brassica family, are grown much like rutabagas, although they mature a good deal earlier. Plant them as a summer crop for bite-sized fruit in about a month, or larger roots two weeks later.

Turnips are an excellent source of vitamins A and C, with good amounts of calcium and iron. Turnip greens are also highly nutritious, and are a prized dish in the South where turnips are fall-sown for a crop of greens all winter. Taste is greatly enhanced by a frost.

We've seldom had a problem with pests or disease, although flea beetles nibble on the greens now and then. If you market the greens as well as the roots, then pay a mind to the beetles. Otherwise you can probably ignore them.

Tip

In the South, you'll find two rules about cooking turnip greens. One, you never just cook greens, you must cook up greens. Second, you never cook up a bunch of greens, or a lot of greens, or even a ton of greens. The only way to cook up turnip greens is to cook up a mess o' greens.

Chapter 14

ORGANIC LIVESTOCK

The best meat I ever tasted, bar none, was from the buck goats I culled and slaughtered every year from our dairy goat herd. These animals were raised on pasture and browse, were able to come and go as they pleased from the barn, were fed only the grain dropped by the milking does, and lived in harmony with us and our land. They were killed humanely, and I did the job myself so they wouldn't know fear.

The best milk I ever tasted was from our Nubian goats, and the best eggs we've ever known were those from our free-range hens, which are fed no growth hormones, no antibiotics, and no other synthetics. They eat our lettuce and other greens now and then, mind you, but we have plenty to go around. Mostly its bugs they're after. Our laying hens are never shut in except on the coldest howling winter nights, and pest

control is left to their summer dust baths. The eggs from these hens have nice bright orange yolks that stand high in the whites and look nothing like their pale super-market cousins.

The goal of organic husbandry today is pretty much to raise livestock and poultry the way I've been doing it for 30 years—in accordance with Mother Nature.

It has to do with raising livestock in the most humane and healthful way possible. My experience is that the healthier the animal, the healthier and more flavorful its milk, eggs, or meat. This means access to high-quality organic pastures, shelter from the elements, and a complete lack of access to virtually any synthetic materials. The routine use of artificial hormones, antibiotics, parasiticides, or other compounds so common in conventional livestock husbandry is prohibited on the organic farm. A sick animal may be treated with antibiotics, but the meat (or in the case of a dairy farm, milk) from that animal can never be sold as certified organic.

As with crop farming, organic livestock begins with the soil. Livestock pastures must be built up organically just as any certified organic grain or vegetable farm must be. This entails the use of compost, cover crops, green manures, and rotational grazing. Pasture seed must be certified organic, and all feed, vitamins, and minerals must be certified organic as well. Just as with crop farms, the livestock farm and its pastures must wait three years before being certified if any conventional inputs, such as fertilizers, were used on the land previously.

The NOP standards require that producers "must manage plant and animal materials to maintain or improve soil organic matter in a manner that does not contribute to contamination of crops, soil, or water by plant nutrients, pathogenic organisms, heavy metals, or residues of prohibited substances."

The healthier the animal, the healthier and more flavorful its milk, eggs, or meat.

Opposite and left:
Livestock, in this case Belted Galway cattle from Scotland, must have access to open, organic pasture land under organic certification rules.

Livestock pastures must be built up organically just as any certified organic grain or vegetable farm must be.

Down the lane at Westwind Farms, an organic meat and poultry operation run by Ralph and Kimberlie Cole.

Put simply, the organic livestock farm not only promotes the health of its own soil and livestock, but the surrounding ecosystem as well. The farm must also be protected from conventional (chemical) spraying used by neighboring farms. NOP requires a buffer to offer protection from chemical drift.

"The edge effect is important," states one organic cattle farmer from Michigan. "Being organic doesn't help much if your neighbor is doing *something* stupid."

Our friends Ralph and Kimberly Cole, who sell their organic meat and poultry in our farmers' market, state their feelings about organics this way:

"We believe good organic livestock husbandry is holistic in nature, striving to promote and enhance the health of our whole environment. This would include not only the animals' health, but also the health of people, their land, and their water.

Good health relies on respect for individual differences and the complexity of nature. To this end, keen observation and monitoring of environmental conditions is an important part of what we do on the farm. We accept that we cannot control our environment (if we ever start believing that, we're making a big mistake!), but we simply strive to work with it for better. Leaving it better and richer than when we received it is our goal."

The key to organic livestock, much like the key to organic produce, lies in the word *sustainable*. This means the farm should be, as much as humanly possible, self-sufficient, with as few off-farm inputs as possible.

In the case of sustainable beef, organic production maximizes the use of pasture forage, and minimizes farm dependence on grain or harvested forage such as silage and hay. Cattle, as ruminant herbivores, act as solar-powered grass harvesters after a fashion, converting a plant material into a

form that humans not only can tolerate, but enjoy. This offers farmers the opportunity to use cattle that are much more efficient grazers than the grain-fed breeds common today.

On our farm at The Hermitage, for example, we graze an old Scottish breed known as Belted Galways, which are accustomed to grazing on some of the most inhospitable land you can imagine. It is said of these incredible gentle beasts, and others like them, that they can put on weight grazing in a parking lot. Some breeds also show heat tolerance and parasite resistance; still others (such as the Galway) tolerate cold well. These and other so-called "minor," or heritage, breeds are often ideally suited to an organic livestock operation.

(I am speaking of cattle here, but the same general organic principles apply to other animals and to poultry. That is, raising them in the most healthful, natural conditions possible.)

With pasture-based beef production, the farmer not only spends less on feed, but can charge more for the meat. On land suitable for row crops, pasture may be included in a crop-rotation plan to interrupt the life cycles of annual weeds and crop pests, and to build the soil by adding manure directly. This is particularly useful on a grain farm where cattle can feed on the grain directly, thereby eliminating the need to sow, fertilize, and harvest the grain mechanically. The only real inputs are land, animals, water, fencing, and good management practices. For this reason, more than a few organic livestock farmers also grow a grain-

Chickens need a place to be safe from predators at night, but require little else in the way of housing.

A rooster (center) isn't necessary for egg production unless you want to raise your own chicks as we've always done.

Budget for 30 percent more winter food than you think you'll need, because in a hard winter extra hay can be hard to find.

crop rotation of corn, soybeans, spelt, oats, and the like.

Trader's Point Creamery, a 100-acre organic dairy within the city limits of Indianapolis, Indiana, feeds its dairy herd no grain at all—nothing but grass. In winter they rely on hay and haylage (or round-bale silage, where forage is baled at a higher moisture content than dry hay and stored in a sealed plastic wrap).

Neil McDonald, the farmer at Trader's Point, feeds his herd nothing but the highest-quality forage, and plenty of it. Replacement cows are raised only on whole milk, and he says this makes for healthy cows that produce years longer than those fed a milk replacement.

"Keep your animals healthy," he says. "Let them build up their immune systems. Here, an ounce of prevention is worth a pound of cure." Whereas many conventional farmers see a sick cow as a problem to be cured, Neil sees a sick cow as a *symptom* of a problem—something wrong with the system. That's the problem he looks for.

Neil advises others who want to adopt a grass-only dairy to budget for 30 percent more winter food than you think you'll need, because in a hard winter extra hay can be hard to find. "Organic dairy farming is growing a lot faster than organic hay and grain farms," he says, so you'll end up paying a premium in January for organic feed.

He also cautions that, while the NOP standards call for a three-year program of soil-building in switching from conventional to organic, "The three years is just a beginning. It can take six or more."

LIVESTOCK MARKETING

The good news is that the market for organic milk, meat, and poultry is growing just as fast as it is for grains and vegetables. Milk in particular is "an exploding market," as one expert put it, where demand is far outpacing supply. As a result, some organic dairies were able (in 2006) to sell at a $4 to $6 premium per hundredweight. When sales are direct to consumer, the premium is even higher (one problem with wholesaling organic milk is transportation costs; the trucks have to drive a long way between organic dairies).

The marketing of organic meat and dairy should emphasize health benefits to both the animal and the environment, according to a 2005 study done in Michigan. Those who said they would pay a premium for such meat and dairy indicated this was because of three things:

The welfare of the animal.

The way it's raised, in a manner beneficial to land stewardship.

The fact that the animal is free of hormones, antibiotics, or other synthetic compounds.

More than 90 percent of the respondents said these factors were important or very important. The fact that the meat was

Cows are just as friendly as goats, and don't get into as much trouble.

raised locally made less difference, but was still important.

One of the first lessons these farmers are learning is not to sell on the hoof to the commodities market. Sell to niche markets, instead. Just remember that to charge more, you not only must have a better product, you must *tell* consumers why it's better. This is not just a sales process; it's a marketing process. It's education. Tell your story, but make only verifiable claims. Tell the truth *always*. Find out and list the Omega 3 value of your milk, for example; don't say it'll make you live longer.

Our friends the Coles direct-market all of their organic Westwind Farm products for on-farm sales, for delivery sale to various locations, at our farmers' market 90 minutes away, and by UPS mail order. They charge a fair price premium, and this is far more than they would receive on an often-volatile wholesale market. Their farm was named Best Small Farm in the state by the Tennessee Department of Agriculture in 2005.

Beginners

In gaining certification for your land, the steps are the same as for crops.

Determine what you'll be raising, and how.

Next, you'll need to select a certifying agent. There were about 60 of them in the United States in 2006, all approved by the USDA. They charge anywhere from about $650 and up, depending on a variety of factors.

Third, you'll complete the application with documentation. This means setting up an organic systems plan, listing all the practices and procedures you'll use, along with all substances used. These will all need to be monitored.

Next is record-keeping, and this turns many against certification but it shouldn't. Any successful farmer keeps good records, and while these may be different than what you're used to, most say it's easy after the first year.

At this point you're ready for the on-site inspection of your farm. Don't worry if the certifier visits and asks for more information; that isn't uncommon.

After this, certification typically takes 90 to 120 days. And you're in business.

We've never had trouble selling fresh eggs from free-range hens, if only because they're so delicious. Don't be afraid to charge more than the most expensive store eggs because yours are fresher.

They had always raised their livestock in a humane and efficient manner, so when national organic standards took effect, about the only difficulty they encountered was finding a supplier of organic seed and feed materials. This was partly due to their location, being far from populated areas.

This is a common problem in the Northeast, where organic livestock farmers have to bring in their organic feed from outside the region. In one University of Vermont study, half the premium earned on the sale of organic milk was burned up in paying more for organic feed. This is less of a problem in the Midwest and east of the corn belt, and it's no problem at all for those who have land enough to (or choose to) raise their own feed.

TRANSITION

Transition to certified organic from conventional livestock is much the same as it is for grain or vegetable farms. You must transition:

The land.

Yourself.

Your markets.

The land requires the same three-year detox and restoration process as other farms, only this time it's for forage pastures. All livestock is required to have access to sunlight, fresh air, and pastures; the exact amount of time they can be housed per year is not a hard-and-fast rule, but is left in the hands of the certifier. For example, a Colorado dairy herd pastured in the shadow of the Rockies may be better off inside much of the year whereas this wouldn't be necessary for a herd in Tennessee.

The application and documentation process, along with the change in your thinking and the change in marketing plans, should put your farm on a higher plane in its relationship to a sustainable earth. And what better way than with livestock.

I know of few things that compare to being present at the birth of farm animals (well, unless it's sheep, in which case your "presence" can be of the up-to-your-elbow variety). Hand-milking is right up there with a birthing.

Slaughtering and butchering, the other side of spring as it were, are no more or less powerful. They're just different—and in this difference lies an intense, overpowering sense of responsibility and honor for all life

---

Certification Checklist

The checklist for a certified organic cattle farm is 25 pages long, but this one page provides a good look at how thorough the certification process is.

INSTRUCTIONS: Complete the following questions to present an overview of the farm and management priorities.

Size of farm (owned)_____

Acreage rented _____

Number of:

Mature cows_____

Replacement heifers_____

Stockers_____

Other types of animal and farm enterprises

_____

_____

Breeds of cattle_____

Number of pastures on farm_____

Number of ponds and water sources _____

Livestock market and months you market in

_____

Months you calve in _____

How many acres of the following do you have?

predominantly cool season perennial grasses_____

predominantly warm season perennial grasses_____

mixture of warm and cool season grasses_____

pastures with legumes_____

cool season annuals_____

warm season annuals_____

pastures that can be stockpiled for late fall/winter grazing_____

FLOWER FARMING

Chapter 15

You'd never think of our friend Peggy Lynn Marchetti as a flower farmer. A singer, sure: She's Loretta Lynn's daughter. An entertainer, yes, for the same reason. A mother, certainly. But a flower farmer? Well, look at her website (www.madisoncreekfarms.com) and you'll see nothing but family and flowers. Yes, a flower farmer indeed.

Peggy has built up her business for more than six years, growing organically in raised beds on the same land her father and grandparents farmed for nearly a century, out of the 1919 farmhouse where her mother wrote songs and filled a gazillion canning jars with winter stores.

"Although we are not certified organic," she says, "we farm within those standards

Per capita, broccoli consumption in the United States has more than doubled in the last 20 years, from 3.5 pounds per person in 1986, to 7.2 pounds per person today.

Peggy Lynn Marchetti on her flower farm.

and have done so over the years, building our soil and keeping our farm and our environment free from harmful pesticides and herbicides while producing top-quality flowers. We are strong supporters of organic farming practices and strive to build a healthy local example of what can be done without poisoning our land and ourselves while preserving our farming heritage and sustainability."

Peggy sells at our farmers' market and at a half-dozen shops in and around Nashville, Tennessee, producing more than 3,500 bouquets a season.

The flowers can't be certified organic because of the chemicals used to preserve them, but organic methods are central to the strength, beauty, and disease- and pest-resistance of her flowers. Our experience over the years has been the same.

Flower farming is a tough row because it can be hard to line up retail outlets, but the grower has the advantage of personal service and a fresh, local product that often outlasts the flowers brought in from outside (or outside the country even). Some sell successfully to restaurants and other places, but be sure the travel is worth your time. A bouquet on every restaurant table is a good thing. You don't want to travel miles, however, to drop off one bouquet at a health spa.

We've had good luck selling flowers with no chemicals or cooler, and having them last a week or more. The trick is to pick with the

An excellent display for cut flower sales. Note the sign to make things human.

sharpest knife possible at the right time of day. For us, that means just after the morning dew has dried, but before the day's heat. Then we'd store them in a shady, airy spot for sale that day and the next. Fertile, organic soil also promotes post-harvest longevity because unstressed plants keep the longest, as do flowers that haven't been over-fertilized.

Zinnias are so easy to grow, it's hard to find a reason not to have some at your farm stand or farmers' market booth.

Fresh 🌸 Designed By You
🌸 Flower Bouquets

MADE TO ORDER

You pick the colors, flowers and fillers that fit your taste and lifestyle. We will make your fresh bouquet on the spot right before your eyes.

LARGE **X LARGE**

$ 10.00 $ 12.00

All signage should be large, easy to read, and concise. But they don't replace the personal touch.

The selection of flowers is entirely up to you (and your florists, if you'll be selling wholesale). My own favorites are zinnias and cosmos, but the list is almost endless with new cultivars coming out every year.

At the farm stand, we'd sell three types of bouquets and label them as such: "Flowers for your bath" (a small nosegay that might dry well), "Flowers for your bedroom" (medium-sized with fragrance), and "Flowers for your table" (large and formal). We also let customers pick their own bouquets from a bucket, or pick right from the gardens.

In addition to flowers, we try to include herbs in a vase. Purple basil and dill are two favorites, but lavender, Genovese basil, and other herbs work. Use your imagination.

Sources: My favorite book on flower farming is *Specialty Cut Flowers* by Allan M. Armitage (Varsity Press, 1993). He does not cover roses, carnations, or chrysanthemums, which comprise the largest part of the world's cut-flower market, but plenty of books and websites can help with advice on these three.

Calendula is grown as a flower, though its petals are considered a medicinal herb for healing cuts.

GLOSSARY

Annual weeds: These are any weeds that flower, produce seed, and then die. Often the cycle is repeated several times in one growing season. The trick is to eliminate them at the flower stage, before they set seed.

Bactericides: Any agent that destroys bacteria.

Bacterial wilt: This disease of cucumber and melon plants is caused by the bacterium *Erwinia tracheiphila*, which invades the vascular (water conducting) tissues in leaves, causing a rapid wilt of the plant.

Broadfork: A two-handled fork up to 30 inches wide with 10 1/2-inch tines; used for deep aeration in compacted soils.

Compost tea: Compost tea is the fluid that results from soaking finished compost in water. It is commonly used to feed seedlings, in transplanting seedlings, and as a foliar feed. There is no strict definition of how much compost per measure of water should be used, and for how long it should soak.

Fish emulsion: This is a high-nitrogen, organic fertilizer made from processed fish wastes.

Cultivar: A cultivar is a named plant variety, selected for some valuable attribute.

Damping-off: A fungal disease that attacks seedlings at the soil level, causing the stem to shrivel and the plant to die. Cool temperatures, moisture, and a lack of air circulation are a welcome mat for this disease.

Eyes: These, in potatoes, are the (usually) sunken, dark spots from which new potatoes grow. In peonies, an eye refers to the swollen, underground growth bud.

Foliar feeding: The act of applying organic fertilizers, such as fish emulsion or kelp, directly to a plants leaves to be absorbed. One quarter teaspoon of liquid detergent per gallon of diluted fertilizer helps keep the mixture from running off.

Green sprout: This is a method of rushing potato production by putting eyes in darkness for several weeks until sprouts appear. Then expose them to at least 8 hours of light daily for two weeks until short, thick, sturdy, green sprouts form. These can be planted outside a week before last frost date.

Harden off: This means exposing a spring transplant to outdoor temperatures gradually, so as not to cause cold shock.

Hot Composting: This is the act of wetting, mixing, and aerating a pile of organic matter such that it heats up to 120 degrees Fahrenheit or more, thus killing most weed seeds and pathogens. Hot compost piles may be turned every few days, and can result in finished compost within a few weeks.

Hundredweight: In the United States, a hundredweight (abbreviated as cwt.) is simply a hundred pounds.

Open-pollinated: This refers to any plant pollinated by wind, insects, or other natural means. These are non-hybrids, and reproduce true from seed, meaning the offspring are virtually identical to the parent plant.

Raised beds: Garden beds whose soil level is raised above that of paths and surrounding ground. This is done to keep soil from compacting.

Rhizones: These are modified plant stems that grow horizontally, under the surface of the soil. Iris "bulbs" are actually rhizomes.

Rootbound: A transplant whose roots encircle the potting medium, the result of growing in a cell or pot too long.

Stolon: A plant shoot that grows horizontally above the ground and produces roots and shoots at the nodes. A good example is strawberries.

Sustainable agriculture: This is a broad term with no single definition, but it generally refers to any farming system that is environmentally sound and relies minimally or not at all on non-farm inputs from outside (such as purchased fertilizer).

Tomato suckers: These are small shoots found growing in the notch between the stem and main branches. These produce relatively little fruit, and are often removed by growers who have the time.

Tubers: a fleshy underground stem or root containing buds from which new shoots arise. The potato is one example.

Zones: This refers to USDA plant hardiness zones, which are used to help determine if winters will be too cold for various plants to survive. These give minimum winter temperatures for every region of the country, but do not tell how long a winter will be, only how cold it can get.

RECIPES

Green beans a lá Greque

In a 3 quart enamel or stainless pot, add 4 cups chicken stock, 1/2 cup olive oil, 2 finely chopped garlic cloves, 1 tsp. ground thyme, 2 tsp. pepper, 1/2 tsp. ground coriander, 1/4 tsp. celery salt, and 1/4 cup lemon juice. Bring mixture to a boil, remove from heat, and let cool about 15 minutes.

Prepare 1 lb. green beans, removing tips and strings, and cutting to desired length. Add the beans to liquid, bring to a boil again, and reduce to simmer until fork tender. Remove beans and let cool. When beans and marinade are cool, combine and refrigerate.

This marinade works well with most any vegetable, so you might want to experiment. You may also cook different vegetables at the same time in the same marinade. Begin with the most delicately flavored vegetables. This makes for quite a visual display and tasty treat for entertaining.

Other vegetables to include are sliced summer squash, cubed eggplant, carrots, artichoke hearts, cauliflower or broccoli florets, diced cucumbers, leeks, mushrooms, pearl onions, or fennel bulb.

Spring onions in Bechamel sauce

Cut the greens off 2 lbs. spring onions (scallions). Reserve these for use in salads, soups, casseroles, or egg dishes. Boil onions until fork tender, drain, and reserve the fluid. Melt 4 tbsp. butter or oil in a medium saucepan. Add 4 tbsp. flour and stir often, about 4 minutes, until the flour is completely cooked (often a light brown color). Add 1 cup chicken stock and 1 cup milk. Stir often until thickened. Add 1/2 tsp. salt, 1/2 tsp. pepper, and 1/4 tsp. nutmeg. Combine sauce and onions in casserole and bake for 1 hour at 350 degrees.

Pizza with fresh Sauce

We have a love-hate relationship with pizza. Love, because it tastes so good, and hate because it's loaded with fat and calories. Let's take another look and see how we can turn this around.

Pizza can and should be one of the most nutritious meals we prepare, and takes very little time.

First, mix 1 packet (tbsp.) yeast in 1 cup warm water. Let this stand about 5 minutes. Add 1 tsp. salt, 1 tsp. sugar, 1 tsp. oil, and combine all together.

Add 1 1/2 cups flour, white or whole wheat. Mix together, then add another 1 1/2 cups flour and kneed until smooth. Divide dough into three balls. Lightly oil three 9-inch pizza pans, then stretch and press dough to fit pans. Don't worry if it pulls apart a bit; just keep working until it submits.

There are endless combinations to use on pizza, but here is an excellent one we like:

Dice at least 1/2 cup (or more) tomatoes for the first layer. Add 1/2 cup peppers, 1/2 cup onions, 1/2 cup sliced mushrooms. Sprinkle on sliced black olives, 1/2 tsp. dried oregano, 1/2 tsp. basil, and salt and pepper to taste. Add 1/2 cup shredded Swiss cheese (or more, but it adds fat calories). Bake at 425 degrees for 25 minutes.

Some of our other favorite pizza additions are diced eggplant, sliced leeks, all types of mushrooms, sliced artichoke hearts, chopped broccoli or cauliflower, or even sliced Jerusalem artichokes. Also water chestnuts, chopped pineapple, and chopped apples. Pesto alone on pizza dough makes an excellent meal.

Sweet Basil Pesto

Basil grows well in nearly every climate. It is so delicious that it should be savored all year long. This recipe for pesto can make that happen.

In a mortar, pound 1 1/2 cups fresh basil leaves (or fine chop with the knife blade in your food processor). Add and chop 2 cloves garlic, and 1/4 cup walnuts. Add 3/4 cup Parmesan cheese. Add 3/4 cup olive oil. Stir all together. Place in jars with lids and refrigerate. The recipe makes about 3 1/3 cups and will stay in your refrigerator almost indefinitely.

Mashed Potatoes

I like to use Russets for mashed potatoes as they generally have a high enough moisture content for the mashing. New potatoes can be really nice as well, and when their skins are nice and thin, I leave them on.

Take four good-sized potatoes (or the equivalent in new potatoes) and cut into about 1-inch cubes. Try to keep a fairly even size. In a 3 quart pot, cover with water and boil at a moderate boil until fork tender (to the cube's center). Drain in a colander and return to heat. Add 6 tbsp. butter and 1 1/2 cups milk or half-and-half. Bring to a good boil and boil for about 1 minute. Remove from heat, add salt and pepper to taste, and whip until nice and fluffy.

Garlic is a great addition. Just add four to six peeled cloves to your potato cubes and boil and mash right along with them.

Easy Broccoli Bake

Cut off the tough end of stem on one good-sized head of broccoli, and cut remainder into bite-sized pieces. Steam or boil until fork-tender, then drain in colander. Assemble broccoli in a casserole dish, drizzle with 4 tbsp. melted butter or olive oil, and the juice of two lemons (about 4 tbsp.). Salt and pepper to taste, sprinkle with seasoned bread crumbs and parmesan cheese. Bake at 350 degrees for 30 minutes. Serves 4 as a side dish.

Janet's Potato Salad

Boil four good-sized russet potatoes until fork-tender (don't let them get too soft). Drain and immerse in cold water. When cool enough to handle, remove skins and chop into bite-sized chunks. Mix with 1/4 to 1/2 cup mayonnaise, depending on moistness of the potatoes. Season with salt and pepper, 1/2 tsp. dried mint (or 1 tsp. fresh mint) and 1/2 tsp. dried oregano (or 1 tsp. fresh oregano). Add 1/2 cup finely chopped sweet onion. Serve on lettuce leaves, garnished with paprika and a sprig of parsley.

Mac and Millie's Cole Slaw

Finely shred one good-sized head of cabbage, two carrots, and one medium onion. Mix together with 1/2 cup mayonnaise (enough to make it creamy), 1/4 cup vinegar, 4 tbsp. cooking oil, 2 tbsp. sugar, salt and pepper to taste.

Cauliflower with Cheddar Sauce

Cut and chop one head of cauliflower and steam or boil until fork tender. Drain (if boiled) and set aside. In a medium saucepan, melt 4 tbsp. butter (or heat 4 tbsp. oil), add 4 tbsp. flour, and cook together over low-medium heat until flour is completely cooked (about 4 minutes). Add 3 cups chicken stock, and let mixture simmer and thicken about 20 minutes. Add 1 cup white wine, milk, or half-and-half. Let simmer again until thick, then add 1/2 cup cheddar or other cheese. Arrange cauliflower in a casserole dish and pour cheese mix over it. Garnish with pepper flakes or parsley. Salt and pepper to taste. Nutmeg, ginger, and cayenne pepper will enhance this dish.

Butternut Squash with Molasses

With its smooth consistency and mildly nutty flavor, butternut squash is my first choice of winter squashes. In a cool, dry spot it will keep for months, and is highly nutritious.

Peel and scoop out the seeds of a good sized squash. Boil squash in a saucepan until fork tender. Remove from heat, drain, and return pan to heat. Add 1 cup milk, 4 tbsp. butter, 1/4 cup molasses, and 1/4 cup sugar (optional). Bring to a second boil for one minute, stirring, then remove from heat, mash, and add 1/2 tsp. ground ginger, 1/2 tsp. nutmeg, 1/4 tsp. cloves, 1/2 tsp. salt and 1/4 tsp. pepper. Mix and serve immediately.

Parsnip (or Turnip) Supreme

I love the distinctive taste of turnips and parsnips, particularly after several frosts when they turn sweeter. They are always a fun, rather exotic, and healthful addition to a meal. First cook six bacon slices until crisp; chop into bits and set aside. Reserve the fat. Peel and chop 2 lbs. of turnips or parships, boil until tender (but not too soft), remove from heat, and drain. Mash and return to low heat and add bacon bits and fat to taste. Add 4 tbsp. butter or oil, 1/4 tsp. ginger, 1/4 tsp. ground coriander, 1/4 tsp. thyme, salt and pepper to taste. Mix well, taste, and adjust seasonings to suit.

Home Dill Pickles

Weigh 3 pounds small pickling cucumbers or thinly-sliced slicing cukes. Pack in clean pint canning jars, leaving 1/4 inch of headroom. Add to each jar: 1 garlic clove, 1 head of dill (or 1 tsp. dill seed) and 1/4 tsp. cayenne pepper. Mix together in stainless or enameled kettle 3 1/4 cups water, 3 1/4 cups vinegar, and 6 tbsp. salt. Bring to a boil, and pour boiling mix over the cukes in the jars (again, leaving 1/4 inch headroom). Tighten lids and process in a boiling water bath for 10 minutes (boiling water should just cover the jars). Makes about 6 pints. The full flavor of these pickles develops in about two to three weeks.

INDEX

About the Author

Peter Fossel has farmed for 30 years, organically for 20.
He was editor of *Country Journal* magazine for years,
and has written for numerous other magazines including
Horticulture, *Organic Gardening*, *American Heritage*,
and the Tractor's Supply quarterly journal, *Out Here*.
He is currently gardens manager at The Hermitage,
home of President Andrew Jackson near Nashville, TN,
selling organic produce through their garden shop.